Defensible Space on the Move

RGS-IBG Book Series

For further information about the series and a full list of published and forthcoming titles please visit www.rgsbookseries.com

Published

Defensible Space on the Move

the Move

Mobilisation in English Housing Policy and Practice

Loretta Lees and Elanor Warwick

WILEY

Registered Office(s)
John Wiley & Sons, Inc., 111 River Street, Hoboken, NJ 07030, USA
John Wiley & Sons Ltd, The Atrium, Southern Gate, Chichester, West Sussex, PO19 8SQ, UK

Editorial Office
9600 Garsington Road, Oxford, OX4 2DQ, UK

For details of our global editorial offices, customer services, and more information about Wiley products visit us at www.wiley.com.

Wiley also publishes its books in a variety of electronic formats and by print-on-demand. Some content that appears in standard print versions of this book may not be available in other formats.

Library of Congress Cataloging-in-Publication Data
Names: Lees, Loretta, author. | Warwick, Elanor, author. | John Wiley &
   Sons, publisher.
Title: Defensible space on the move : mobilisation in English housing
   policy and practice / Loretta Lees and Elanor Warwick.
Description: Hoboken, NJ : John Wiley & Sons, 2022. | Includes
   bibliographical references and index. | Contents: Defensible space: an
   introduction -- Defensible space is mobilised in the UK -- Defensible
   space goes on trial but attracts those in power -- Operationalising
   defensible space -- Evaluations of defensible space -- The uptake and
   resilience of defensible space ideas -- Defensible space: a common
   sense, middle-range theory.
Identifiers: LCCN 2021044849 (print) | LCCN 2021044850 (ebook) | ISBN
   9781119500445 (hardback) | ISBN 9781119500438 (paperback) | ISBN
   9781119500414 (pdf) | ISBN 9781119500407 (epub) | ISBN 9781119500421
   (ebook)
Subjects: LCSH: Housing policy--England. | Crime prevention and
   architectural design--England. | Residential mobility--England. | City
   planning--England.
Classification: LCC HD7334.E53 L33 2022 (print) | LCC HD7334.E53 (ebook)
   | DDC 363.5/5610942--dc23/eng/20211208
LC record available at https://lccn.loc.gov/2021044849
LC ebook record available at https://lccn.loc.gov/2021044850

Cover image: Balfron Tower/Brownfield Estate 2014 © Michael Mulcahy
Cover design by Wiley

Set in 10/12pt PlantinStd by Integra Software Services, Pondicherry, India
Printed and bound by CPI Group (UK) Ltd, Croydon, CR0 4YY

The information, practices and views in this book are those of the author(s) and do not necessarily reflect the opinion of the Royal Geographical Society (with IBG).

C088735_260122

Contents

List of Figures

List of Tables

Glossary of Acronyms

ACPO	Association of Chief Police Officers
AEDAS	An international architecture and design practice
ALO	Architecture Liaison Officer (now DOCO)
BSI	British Standards Institute
BS	British Standard
CABE	Commission for Architecture and the Built Environment
CPDA	Crime Prevention Design Advisor (now DOCO)
CPTED	Crime Prevention Through Environmental Design
DICE	Design Improvement Controlled Experiment
DCLG	Department for Communities and Local Government
DfT	Department for Transport (now DTLR)
DETR	Department for Environment Transport and the Regions
DLUHC	Department for Levelling Up, Housing and Communities
DoE	Department of the Environment
DOCO	Designing-Out Crime Officer
EAP	Estates Action Programme
GLA	Greater London Authority
GLC	Greater London Council
HAT	Housing Action Trust
HCA	Homes and Community Agency (now Homes England)
HDD	Housing Development Directorate
HIP	Housing Investment Programme
HO	Home Office
HTA	Hunt Thompson Associates Architects
HUD	United States Department of Housing and Urban Development
KCL	King's College London
LSE	London School of Economics
LSVT	Large-Scale Voluntary Transfer

MHCLG	Ministry of Housing Communities and Local Government (now DLUHC)
MORI	UK market research company – now Ipsos MORI
NACRO	National Association for the Care and Resettlement of Offenders
NDC	New Deal for Communities
NPPF	National Planning Policy Framework
ODPM	Office of the Deputy Prime Minister
PEP	Priority Estates Project
RIBA	Royal Institute of British Architects
SBD	Secured by Design
SNU	Safe Neighbourhoods Unit
SRB	Single Regeneration Budget
SRD	Social Research Division
UCL	University College London
UHRU	Urban Housing Renewal Unit

Series Editors' Preface

The RGS-IBG Book Series only publishes work of the highest international standing. Its emphasis is on distinctive new developments in human and physical geography, although it is also open to contributions from cognate disciplines whose interests overlap with those of geographers. The Series places strong emphasis on theoretically informed and empirically strong texts. Reflecting the vibrant and diverse theoretical and empirical agendas that characterise the contemporary discipline, contributions are expected to inform, challenge and stimulate the reader. Overall, the RGS-IBG Book Series seeks to promote scholarly publications that leave an intellectual mark and change the way readers think about particular issues, methods or theories.

For details on how to submit a proposal please visit:
www.rgsbookseries.com

Ruth Craggs, *King's College London, UK*
Chih Yuan Woon, *National University of Singapore*
RGS-IBG Book Series Editors

David Featherstone
University of Glasgow, UK
RGS-IBG Book Series Editor (2015–2019)

Acknowledgements

We have so many people we would like to thank, not least of whom is Alice Coleman for opening up her archive to us and for answering our questions. To Wiley and the editors of the RGS-IBG book series for their patience as Loretta recovered from surgery on a torn rotator cuff and Elanor a torn calf muscle during the initial writing of this book – the craziness of us both working on the book on crutches and with an immobilised arm was a sight to behold! And then whilst undertaking revisions during Covid, both of us with children home doing online learning. To Jane Jacobs now working in Singapore who shared our initial interest in defensible space. To all the interviewees in the book who kindly shared their time, memories and thoughts. To the many professional colleagues from organisations known by acronyms (ACPO, EDGE, GLA, LBTH/THH, MHCLG [now DLUHC], Poplar HARCA, CABE) who opened doors and passed on contacts. To Simon Harding who gave Loretta the United Kingdom's first (until Grenfell) national tower block survey from the late 1980s/early 1990s plus archival material. To Kopper Newman, Oscar Newman's wife and estate executor, for permission to reprint his images as long as he is given due credit. To ESRC grant ES/N015053\1 on council estate renewal in London that Loretta was Principal Investigator on. And last but not least, to our families who have supported us through the research and writing of this book: Rob and Dora (Elanor); David, Meg and Alice (Loretta). Thank you all.

Preface

Defensible space was one of the first spatial debates Elanor was consciously aware of as a young architectural undergraduate at the Bartlett. The arguments following the publication of *Utopia on Trial* in 1985 (and Bill Hillier's critical lectures on spatial analysis logically unpicking the dissonance between theory and application) stirred Elanor's awareness of the extent that architectural design decisions really affect people's lives and experiences. The notion of differentiated public and private spaces remained formative throughout Elanor's architectural and urban-design practice, and later professional jobs researching housing design and policy. Directing a Home Office funded study into design and crime on housing estates, provoked a sense of déjà vu. Hadn't Alice Coleman been asking similar questions about natural surveillance and symbolic ownership of public areas 20 years previously? Were the spaces around these award-winning high-density new homes at risk of becoming as unloved as the decrepit and run-down brutalist concrete estates Coleman had studied? Elanor's PhD on defensible space (as a geographer not an urban designer) was motivated in part by this sense of long-unresolved questions and a fear of repeating past design mistakes. So, when Elanor finally met Alice Coleman, at her 90th birthday celebration at King's College London, surrounded by ex-students and academics applauding her research successes and being King's first female professor of geography, their conversation about the contradictory nature of defensible space was lively and questioning.

Loretta first came across the concept of 'defensible space', age 17, in her A-level History of Art module 'Design for Living', which looked at high-rise architecture. She went on to research her A-level Geography project on defensible space in high-rise blocks in the New Lodge area of Belfast in Northern Ireland, but had to hire (paid for with a box of chocolates) a school friend (who lived in one of the blocks) to do the survey work, as she could not go herself into the high-rises in what was a 'no-go area' of Catholic Belfast with an English accent. On taking up her first permanent academic position in Geography at King's College London in 1997 she met Alice Coleman – then an emeritus professor – face to face for the first time, and found her to be most welcoming. She next encountered the concept of 'defensible space' when supervising an ESRC-ODPM funded project on high-rise living in London in which the evidence base on high-rise living was reviewed. Loretta later teamed up with Jane Jacobs who was then working at Edinburgh University on the Red Road high-rises in Glasgow, and they both interviewed and videoed Alice Coleman. Loretta also first supervised Elanor's PhD, part funded by CABE, on defensible space. Defensible space reared its head yet again when Loretta started researching the demolition (and gentrification) of the Heygate and Aylesbury Estates in London, and, as mentioned in the book, the lack of defensible space was used by Southwark Council to argue that the Aylesbury Estate needed to be demolished at the 2018 public inquiry in which she was an expert witness.

It is clear that defensible space's relevance remains. Responding recently to an evidence call from MHCLG on the impact of segregation of communal spaces on mixed tenure blocks, Elanor reiterated the negative impact of confused territoriality, blurred private space and divisive 'poor doors' that continue to be built into new housing schemes. For such a familiar and easily recognised concept, defensible space is still deeply misunderstood. This is why we wanted to write this book for academic researchers *and* practitioner colleagues across all geographic and built-environment disciplines, providing a practical example of how to apply this research learning in the future, as much as a historical record of the evolution of the concept of defensible space.

Chapter One
Defensible Space: An Introduction

in the worst estates…you're confronted by concrete slabs dropped from on high, brutal high-rise towers and dark alleyways that are a gift to criminals and drug dealers. The police often talk about the importance of designing out crime, but these estates actually designed it in. (British ex-Prime Minister, David Cameron, *The Sunday Times*, 2016)[1]

Social scientific knowledge linking environment and behavior precipitated the British shift away from public housing and was used to promote several types of privatization. (Cupers 2016: 183)

'Defensible space' is a highly contested concept and approach to designing out crime, frequently applied to public housing estates in the United Kingdom, North America, Europe and beyond. It is both an urban idea and a policy concept, arguably the most influential concept in built environment crime prevention to date. In this book we use 'defensible space' as a vehicle to explore how movement/mobility/mobilisation (and we discuss how these three are related but different later in this chapter) changes ideas/concepts. In exploring the movement/mobility/mobilisation of defensible space from the United States to the United Kingdom and into English housing policy *and* practice[2] we extend recent work in geography, and indeed urban studies and urban planning more widely, on policy mobilities in a number of critical ways.

The idea of defensible space was introduced to the United Kingdom through a book by North American architect/planner Oscar Newman and a 1974 BBC Horizon television programme on his ideas. Our book traces in detail the

Defensible Space on the Move: Mobilisation in English Housing Policy and Practice, First Edition. Loretta Lees and Elanor Warwick.
© 2022 Royal Geographical Society (with the Institute of British Geographers). Published 2022 by John Wiley & Sons Ltd.

dispersal/embedding of the concept of 'defensible space' in England from the 1980s onwards from the point where geographer Alice Coleman reintroduced and popularised it in the English context. For this we revisit Coleman's critique of modernist council high-rises in England in her 1985 book *Utopia on Trial*, in which she outlines her conceptual (which she hoped to operationalise) account of defensible space. We look in detail at her research and the sometimes quite vicious criticisms of it from other geographers, architects and planners. We use in-depth interviews and oral histories with Coleman herself, and other housing researchers and practitioners from the time, to piece together the story of how this geographer took Prince Charles on a field trip to look at the problem of defensible space on a public housing estate in London, and how she managed to get a one-to-one meeting with Margaret Thatcher, persuading the then Prime Minister to give her £50 million to pilot her ideas for retrofitting council estates with defensible space principles.

We discuss the pilot projects themselves, moving on from Coleman's conceptual treatise to an operational account of defensible space as demonstrated through her Design Improvement Controlled Experiment (DICE), and how this influenced the wider context of English housing policy and practice at the time. The book explores the multiple ways the concept of defensible space was interpreted and implemented, as it circulated from national to local level and within particular English, especially London, housing estates; illustrating how the transfer mechanisms worked at both a policy and practitioner level. Despite being a concept whose principles continue to underpin design guidance (such as Secured by Design [SBD]), defensible space failed to coalesce into a single formal policy, remaining a cluster of associated disputed elements. How these conceptual elements aided or hindered transfer and take-up of the concept is noted by tracking routes to acceptance, the roles of formal transfer mechanisms, informal information sharing by transfer agents traversing networks, or practitioners' local contextualisation of generic guidance. Our research demonstrates the ongoing resilience and acceptance of 'defensible space' from the 1970s into the 2000s, despite multiple criticisms of architectural/environmental determinism, of being unproven scientifically, and the vague and inchoate nature of the concept. More recently, though, there is evidence that defensible space is beginning to be erased, and expunged, from planning and urban regeneration policy. Nevertheless, we argue for greater trust in practitioner experience and on the basis of its continued usage, that defensible space is positively ambiguous, it has neither been proven nor disproven, and as such is a middle-range theory: 'between the minor hypotheses of day to day research and unified theory' (Merton 1967: 39).

Following Flyvbjerg (2001) we show how the simplified dualisms of theory in academia are helpful for polemic thinking and writing but they 'inhibit understanding by implying a certain neatness that is rarely found in real life' (p. 49).

Flyvbjerg notes that policy makers get around this messiness by pragmatically asking: 'will this solution work here?' Leaping to a solution is different from understanding a theory, and McCann (2008) is disappointed by the 'paucity of detailed critical geography knowledge of how policy making works' (p. 4). Our book provides much needed insight into this, and in doing so expels some of the myths that good social science will follow a straightforward route into policy. We develop Jacobs and Lees' (2013) earlier account of defensible space on the move based on three further insights: a) that policy does not move as an homogeneous, fully formed piece, but as disaggregated elements (of pre-policy, sub-policy epistemes or practices); b) these fragments of knowledge are translated into policy only in context (in situ); c) that the relationship between academic research and practice is not a simple linear progression of policy appropriating and utilising university created research. The interplay between academic knowledge and policy we describe is complex, contingent and often controversial.

Peck and Theodore (2010a) recognise that policy transfer is often disrupted by the messy realities of policy making at the ground level, yet little reference is made in the existing policy mobilities literature to the messy realities of practice. Although recent reviews, like those discussed later in this chapter, provide a very useful overview of (conceptual) evolution in the field of policy transfer/mobilities, 'they do not provide an overall explanation of policy transfer processes and outcomes' (Minkman et al. 2018: 224). This book fills that gap by looking at how defensible space was put into practice in England, addressing McGuirk's (2016: 94) request for research on 'the "how" questions of practice'. Much of the policy mobilities literature also only follows one or two mechanisms of transfer; in this book we follow a dozen or so mechanisms, showing a far messier and more interconnected reality than the literature suggests. In doing so we elaborate on the challenges of tracing power and the role evidence plays within the policy making process. This has lessons for the utilitarian turn in social research that happened in the 1990s. We argue that to some degree it does not matter if defensible space is a fundamentally poor idea/concept (which we discuss in terms of both its strengths and weaknesses); what is more important is its mis/match with policy contexts or success/failure due to the personalities involved. That defensible space has moved into the mainstream, without definitive proof or consistent government support, is due in no small part to Coleman's geographical work. Despite the uncertainty surrounding it, defensible space continues to be promoted as a powerful and influential way of salvaging so called 'sink' estates, as the former Prime Minister David Cameron, like other Prime Ministers before him, called them. The ideology of contemporary estate demolition in the United Kingdom has drawn heavily on the US Federal Department of Housing and Urban Development's HOPE VI programme of public housing demolition and renewal which itself draws on defensible space principles to rectify the problems of past failed projects (see Popkin et al. 2004).

In the round, this book makes an important conceptual contribution to policy mobilities thinking, but also to policy and practice, explaining practitioners' handling of complex spatial concepts, through the practical application of an idea that is, as we show, a middle-range theory. We also use a primary transfer agent (a geographer – Alice Coleman) and a concept (defensible space) to reflect on the role and contribution of British geography in English housing/planning policy. Our conceptual framework looks at positionality, context, multiple perspectives, ambiguity and mutability. The irony being that Coleman's positivist, non-negotiable view would totally reject this interpretation as too complex. Although Coleman is the primary transfer agent, we also discuss attendant ones, including one influential individual who acts as a foil to Coleman's views. Nonetheless, Coleman is a useful prompt through which to explore the conflated cluster of sometimes contradictory concepts that are gathered together under the umbrella of defensible space. As an unusual geographer and scholar, known not only for her eclecticism – in Maddrell's (2009) view 'a polymath generalist'– but also for her outspoken views and right wing politics, it is remarkable the extent to which her radical view was applied consistently and rigorously. Like herself, Coleman's take on defensible space was uncompromising, rather than fluid and negotiated. Coleman's values were outliers in the wider discipline of geography at the time (and remain so today), yet they impacted the canon of planning and urban design from another perspective, and continue to do so in recent debates over the demolition and refurbishment of council estates in the United Kingdom.

The Origins of Defensible Space

At its simplest, defensible space can be defined as 'space over which the occupiers of adjacent buildings can exercise effective supervision and control' (Cowan 2005: 102). Given it is about control over space it is an inherently geographical concept. The notion of defensible space demonstrates the interrelationship between the physical design of spaces, social interaction and crime. Yet the concept of defensible space is contradictory, tentative and ill defined. Indeed, the concept has proved ambiguous and malleable enough to support diverse interdisciplinary interpretations; it is a ubiquitous, familiar idea, not only to built environment professionals but also to academic geographers, criminologists, architects, and so on. Defensible space remains a 'common-sense concept' recognised even by the general public in its most basic form.

Architect-planner Oscar Newman (1935–2004) is often said to have coined the term 'defensible space', although it was first used by the sociologist William Yancey. In 1972 Newman published *Defensible Space: Crime Prevention through Urban Design*, based on his research on New York City public housing. He later refined his thesis in his, 1976 *Design Guidelines for Creating Defensible Space*. But the emergence of the notion of defensible space has a longer history and is far richer and more complex than Newman's method of (re)designing spaces to inhibit criminal activity. Urbanist Jane Jacobs (1961) had famously talked about the importance of 'eyes on the street' sometime before, and we can trace the origins of the concept under different names within Lewis Mumford's (1938) *The Culture of Cities*, or in William Whyte (1956) and Kevin Lynch's (1960) observations. Concepts rarely, if ever, have a singular origin and defensible space is no different. The basis of defensible space emerges from notions of social interaction and encounter in modern city streets. Mumford, who at one level was very confident about the link between civility and the city, was also concerned that the emergence of the 'megalopolis' was leading to 'anonymity' and 'impersonality', which he saw as 'positive encouragement to asocial or anti-social actions' (Mumford 1938: 266; see also Fyfe et al. 2006). The idea of defensible space also builds on the historical work of Louis Wirth and George Simmel and other urban sociologists exploring the influence of the form and character of the modern city (see Lees 2004), or the interconnectedness of common space on the experiences of its inhabitants. Whyte's (1956) study of urban public spaces demonstrated the complex positioning of privately owned public spaces; he studied the behaviour of people within small New York places, and started to catalogue the successful and unsuccessful elements of such spaces. His observations on 'people moving' and 'people watching' in plazas, fed through to Jacobs' (1961) descriptions of 'ballet on the street' and 'eyes on the street'.

Most commentators agree that defensible space is not solely about crime, nor is it limited to the design of spaces and fences, the physical location of windows, or the layout of streets and neighbourhoods. Obviously spatial perspectives on the relationship between crime and housing have existed since Booth's nineteenth century maps of poverty and social class in London, Beames' studies of British rookeries or Burgess' pre-First World War concentric zone model of Chicago. These neighbourhood analyses established purely spatial patterns of crime, lacking behavioural differentiation. Basic, area-based approaches evolved into the ecological analysis of the impact of poor housing, poverty and transient populations during the 1950s. These more nuanced methods maintained that a particular behavioural setting had the power to elicit similar responses from diverse occupants; yet remained at an aggregate inter-urban scale of investigation. Crucially, by relating image, meaning and legibility to perceptions of place, Lynch (1960) transformed how designers and social scientists alike perceived urban form. Jones and Evans (2008: 115) point out that 'the anti-modernist stance' of defensible space ideas emerged from Lynch's (1960) notion of legibility, and his criticism of urban spaces where movement and function was not clear. Later, Lynch's (1981) articulation of notions around spatial rights – 'right of presence', 'use and action', 'appropriation' and 'modification' (p. 205) – introduced designers to concepts integral to ownership and control of spaces, fostering freedom to engage with others or being free to retreat from threat.

Jacobs articulated a planning view of defensible space as it is experienced on city streets. In *The Death and Life of Great American Cities* (1961) her combination of social science and planning theories presented a positive vision of urban living, where active street life and numerous social interactions were identified as indicators of successful, well-designed places. She noted that more crimes occurred in the often-deserted public spaces found in (modernist) public-housing projects, than in traditional, crowded streets. Poyner (1983), amongst others, identified Jacobs' influence on Newman's idea of defensible space; her other critical legacy was a taxonomy of space that shaped social interaction. Jacobs (1961) succinct definition of the three attributes of safe spaces included a plea for clear delineation of public from private space:

> First there must be a clear demarcation between what is public space and what is private space. Public and private spaces cannot ooze into each other as they do typically in suburban settings or in projects.
>
> Second, there must be eyes on the street: eyes belonging to those we might call the natural proprietors of the street. The buildings on a street equipped to handle strangers and to ensure the safety of both residents and strangers must be orientated to the street.
>
> And third, the sidewalk must have users on it fairly continuously both to add to the number of effective eyes on the street and to induce people in buildings along the street to watch the sidewalks in sufficient numbers. (Jacobs 1961: 35)

Newman's version of defensible space (outlined in the section 'Oscar Newman's Defensible Space') extended Jacobs' two-part definition of public and private space into four parts: private, semi-private, public and semi-public space.[3] Inevitably, defensive architecture for Jacobs, but especially Newman, was connected to post-Watergate and Vietnam progressive US ideals for personal, social and economic liberation; and the city was an emancipatory site for these beliefs. As Newman (1972: 203) said: 'For our low-income population, security in their residential environment – security from the natural elements, from criminals and from authority – is the first essential step to liberation'. At that moment in time, in the early 1970s, there was a concern for the poor, welfarism and community commons; progressive design was linked to social emancipation. One political contraction was the rapid move from emancipatory concern into a strategy of control under the revanchist city and the shifting political climate in US cities like New York (see Smith 1996, on zero tolerance and the revanchist city).

Similarly, in the early 1970s criminological research moved from mapping the location of offenders or offences to proposing models of the local environment as a framework for individual behaviour. Yancey (1971) suggested a hyper-local explanation for residents' actions. His influential paper on the Pruitt-Igoe public-housing project in St Louis, Missouri (which had become infamous internationally for its crime, poverty and racial segregation) critiqued the project's design, particularly its lack of semi-public space which housing management professionals considered to be unnecessary 'wasted space' (Yancey 1971: 11). But far from being wasteful, Yancey argued that lacking this shared space resulted in residents retreating into private internal spaces. Yancey termed these semi-private areas 'defensible space' (1971: 17) using the term several years before Newman. He found that residents easily recognised the difference between private apartment space and public shared amenities. Similar to residents socialising on their front steps (or stoops) in the North End of Boston, as described by Jacobs (1961), Yancey (1971) noticed that the semi-public space outside family homes 'provides the ecological basis around which informal networks of friends and relatives may develop' (p. 17). These spatial relationships shaped the social networks of residents, with the constrained physical design limiting occupants' interactions beyond the confines of their individual homes. The design of our surroundings then came to be seen as a fundamental method of demarcating private territory physically or symbolically (Ley 1977; Lynch 1960; Sennett 1986).

More recently, research into the geography of the fear of crime describes a subtle and complex sequence of interactions that link the design of the physical environment to individuals' perceived potential for crime, and hence to an individual's sense of wellbeing (Smith 1986a, 1987, 2003). Rudlin (2015), for example, considering the proliferation of gates onto courtyards and walls forming inward-facing gated communities, points out that the creation of defensible urban forms should not only be able to make residents feel safer, but contribute to making the surrounding city safer. He is concerned that gates and barriers result in occupants 'only mixing with like people, losing their ability to live within a diverse society and increasing their level of fear, thus fuelling a vicious circle in which they feel the need for more protection and control' (Rudlin 2015: 37–39). Blandy (2007: 47) is especially critical:

> gated communities are not an effective response to current issues of crime and disorder in terms of physical security and collective efficacy; nor do they assist in regenerating deprived areas, or tackling problems of disorder on large social rented estates. Indeed, any further growth in the collective fortification of affluent homes and retro-gating of social rented estates is likely to contribute to increased social divisiveness.

So – too few people on the street, not enough movement or activity and public space, were seen as pathological. Modernist planning and social housing projects that lacked activity engendered failed public space, their segregation from active street networks explaining their abnormality – all prompting calls for 'defensible space'.

Oscar Newman's Defensible Space

Newman's own career in housing research, like Yancey, also started in St Louis where he taught architecture and city planning at Washington University from 1965.

He wrote about watching the decline of Pruitt-Igoe, triggering his recognition of Yancey's novel concept and prompting his development of the idea of defensible space (Newman 1995). Newman conceptualised defensible space as a combination of spatial and social mechanisms, with the capacity to create physical zones of territorial influence, to provide natural surveillance opportunities for residents and to positively affect the (often negative) perception of a public housing scheme's distinctiveness and resultant social or economic stigma (see Figure 1.1). Newman refined this definition in his *Design Guidelines for Creating Defensible Space* as:

> a residential environment whose physical characteristics—building layout and site plan—function to allow inhabitants themselves to become key agents in ensuring their security. (Newman 1976: 4)

Newman's own research was undertaken in New York City where he studied two housing projects, drawing on New York Housing Authority Police crime statistics, as well as his own data generated through resident interviews and building analysis. One housing project with high-rise blocks, Van Dyke, had a 50% higher crime rate than the other, Brownsville, which consisted of mid-rise walk-ups (see Figure 1.2). Finding higher crime rates in the lifts, stairways and landings of the high-rise apartment buildings, Newman argued that the Van Dyke residents felt no personal responsibility for the communal areas shared by many occupants. Good design and certain physical characteristics, he argued, allowed Brownsville residents to monitor and occupy semi-private spaces, ensuring their security. Although intended as practical, applied research, Newman's assertions about verifiable scientific methods enhanced the reputation of his work (he maintained the study compared identical communities, with constant

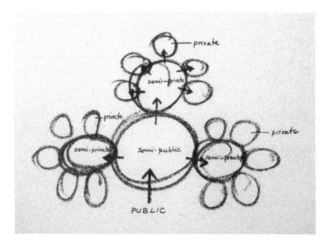

Figure 1.1   Newman's typology of space. (Permission: Kopper Newman)

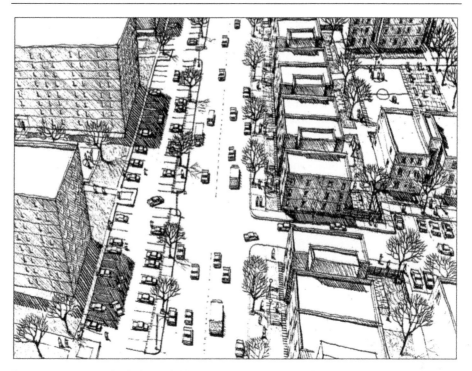

Figure 1.2   Newman's sketch of Van Dyke (L) and Brownsville (R), New York City. (Permission: Kopper Newman)

social characteristics, only the building forms varied). However, these positivist methodological assertions legitimised later scientifically based criticisms of Newman's research.

Newman's concept of defensible space as derived from these studies elaborated on Jacobs' attributes of safe public spaces. Safe spaces required: 'visibility to witnesses, community spirit and being prepared to guard neutral territory, a stream of potential witnesses and demarcation of private territory physically and symbolically' (Mawby 1977: 171). The 'dimensions' that underpinned Newman's defensible space concept were:

- the capacity of the physical environment to create physical zones of territorial influence,
- the capacity of physical design to provide surveillance opportunities for residents and their agents,
- the capacity of physical design to influence the perception of a project's uniqueness, isolation and stigma, and
- the influence of geographical juxtaposition with 'safe zones' on the security of adjacent areas (Newman 1972: 50).

Territoriality can be interpreted very broadly from the concept of place attachment or social commitment to location, to animal-like defending of 'turf', through to symbolic or physical acts. Similarly, territoriality can be applied at a range of scales, relating to specific sites or wider neighbourhoods. The positive and negative attributes of territoriality have been explored across the social science disciplines; a contradictory interpretation of territoriality arises from interconnected mechanisms that establish control over space. This control can be communicated in a multitude of ways: physical (a person's presence reducing and inhibiting criminal activity or a physical barrier such as fences or walls); symbolic (plants and/or other symbolising ownership of the space); visual (graffiti tags); or physiological (encouraging a sense of safety through well-lit, well-looked-after places). Control mechanisms applied to the same space that communicates possession can result in alternative readings from a range of users. To Newman, territoriality consisted of several interconnected effects: it was a process of establishing a sense of ownership for legitimate users of the space, which in turn provided a clear definition of areas controlled and influenced by inhabitants, and was encouraged through familiarity with neighbours or passers-by.

Territoriality is often characterised as 'the absence of anonymity' for legitimate inhabitants. Ley (1974a), for example, emphasised opportunities for casual contact, leading to joint ownership of space and to the exclusion of non-inhabitants. For Newman, territoriality was directed at strangers by controlling their access and movement through a housing estate or a neighbourhood. Yet, territoriality might influence factors that jeopardise as well as improve security. The number or frequency of passers-by has contradictory interpretations, providing a spectrum of explanations; many strangers might provide greater anonymity for criminals, but the same number of familiar passers-by might act as potential witnesses or provide positive opportunities for social interaction.

Surveillance is especially emphasised in the concept of defensible space, which favours the use of natural or informal surveillance (windows overlooking public spaces, or active street frontages) over mechanical (CCTV) or formal forms of supervision (security guards). Natural surveillance can increase a sense of security, which will encourage the greater use of a space and positively reinforce ownership. However, even the apparently simple tactic of natural surveillance can have contradictory consequences on design decisions – ground floor windows, which may improve visibility over adjacent spaces, might also advertise opportunities for burglary.

Image is the least well-defined, but in some ways the most tangible of Newman's dimensions; he includes the use of distinctive built forms, materials, finishes or aesthetics, which are identified as having associations with a particular social class or lifestyle. Image and its resultant associations suffuse housing design from the urban scale to the detail, from the size and scale of the blocks themselves to the selection of particularly institutional or utilitarian materials. Image is, of course, affected by the physical condition of the buildings and spaces; by signs of decay, neglect or poor maintenance. The dominance of materials in the reading of spaces is widespread: cracked fluorescent lights cruelly exposing the flaws in cold damp concrete ceilings, speak of tight budgets, lack of care and neglected maintenance; the aesthetic of 'cheap'n'drab' has become synonymous with public housing stigmatised by its very appearance and design. Such images became encapsulated in the broken-windows theory discussed later.

Milieu is the positive influence of activities perceived as safe (such as police stations or well-used streets) adjacent to areas suffering from crime. Greater levels of activity contribute to a positive milieu through a potentially benevolent increase in passers-by. It encompasses the effects of neighbourhood context and reputation, and identifies urban design decisions for the location of amenities as an explanation for spatial concentrations of crime or the presence of unpopular areas. Certain activities or amenities (shops, clubs, bars or even neglected recreational areas) act as attractors for crime. Newman (1972: 108–109) explained that 'certain sections and arteries of a city have come to be recognized as being safe – by the nature of the activities located there; by the quality of formal patrolling; by the number of users and extent of their felt responsibility; and by the responsibility assumed by employees of bordering institutions and establishments'. Here, Newman (1980) touched on the Chicago School ideas of Burgess, Shaw and McKay of crime hot spots through mapping high- and low-risk neighbourhoods. He ascribed a simplistic juxtaposition of high crime and low crime areas and proposed pepper-potting potentially difficult residents within quiet well-behaved neighbourhoods to reduce crime levels. To test his hypothesis, Newman analysed and redesigned Five Oaks in Dayton, Ohio, into mini-neighbourhoods by removing through traffic, changing the character of streets to enable children's play, increase neighbours' interactions and recognition of cars, and altering the layout so potential criminals had to enter and exit one way. Newman countered complaints that these changes had displaced crime by arguing that positive milieu, that is the residents of Five Oaks taking control of their streets, extended out into the bordering communities as well. Nevertheless, milieu is an ambiguous mechanism relating to spatial layout and land-use patterns that provides a context for routine[4] activities (Reynald and Elffers 2009). Habraken's (1998) tripartite ordering of the physical, territorial and cultural readings of spaces into a 'common fabric' reminds us of the interrelated nature of these characteristics. It is the interaction of image, milieu and natural surveillance, which strengthens or weakens territoriality or pride in a place.

Before turning to criticisms of Newman's defensible space theory, the racialised contexts of Newman's writings should be noted. Notoriously Pruitt-Igoe, Newman's trigger for developing his theory of defensible space, was a dumping ground for over 10,000 low-income African Americans. It is important to situate defensible space in relation to the way that discourses of crime in the United States were (and continue to be) profoundly racialised. Racial inequalities had prompted race riots from the late 1960s in cities like New York, Los Angeles,

Newark and Detroit, leading to federal concerns about crimes against property and the pathological construction of black (male) (inner-city) bodies as criminal and dangerous (see Smith 1996). The photographs in the federal reports commissioned at the time, according to Knoblauch[5] 'put a face to the so-called pathology of the "personality factory" of black families in the "slums"', and 'displaced concern with structural economic issues onto residents themselves'. Newman saw this de facto segregation as negative, and in the face of waning political support in federal programmes to address such issues he saw defensible space as a physical solution, but he also stridently criticised welfare dependence. As Kinder (2016: ch. 3) points out:

> Newman's focus on physical design and social exclusions was problematic. His work undertheorized criminality and inequality. He explicitly discouraged city planners and policy makers from using social, economic, and welfare planning to revitalize cities, create jobs, or redistribute income.

The idea that Pruitt-Igoe's design, and not structural inequality and racism, was to blame took hold, and Newman's ideas about territoriality and defensible space became mainstream in assumptions about public housing redesign in the United States. Indeed, they heavily influenced The US Department of Housing and Urban Development's (HUD) HOPE VI program of demolition and renewal of public-housing projects across the United States (see Vale 2013).

Criticisms and Endorsements from Different Disciplines

Newman's concept of 'defensible space' was mobilised from the United States to the United Kingdom (see Chapter 2) at the same time as vociferous cross-disciplinary critique of it emerged. Critics implied that Newman had manipulatively selected which housing projects to examine (Bottoms 1974; Hillier 1973; Mawby 1977; Merry 1981); for while he mentioned several positive and negative examples of neighbourhoods and projects, he only talked about a single pair of housing projects (Van Dyke and Brownsville) in any detail. Critics argued he failed to demonstrate the comparability of these two projects to other examples; of course, finding truly 'paired' public housing projects to compare may be impossible but the differences that Newman ignored were ones that influenced his arguments greatly. For example, Brownsville was completed over a decade earlier than Van Dyke, so it was perhaps further advanced along a cycle of popularity or decay (Bottoms 1974). Others pointed to Newman's shallow ethnographic summary of resident interviews as methodologically weak (e.g. Ley 1974a). In addition, when characterising the projects as having similar resident populations in terms of income, race and family size, Newman ignored the less favourable reputation of one project that may have discouraged better-off residents; as such he was accused of omitting potentially conflicting or explanatory data on education or social class (Hillier 1973; Mawby 1977).

Researchers also identified statistical flaws in Newman's data analysis, questioning both the selective use and underplaying of statistics (Bottoms 1974; Ley 1974a; Mawby 1977). Newman failed to separate out strength of correlation or causation for his proposed linkages between crime levels and built form. Newman (see also Coleman in Chapter 2) concentrated on measuring easily quantifiable effects (Moran and Dolphin 1986), relying on statistical analysis of numerical data (Coleman 1985a; Newman 1972). Subsequent defensible space research applied more rigorous and in-depth methods, for example, Merry's (1981) detailed ethnographical victimisation studies that significantly improved on Newman's anecdotal stories, or Cozens et al.'s (2001a) exploration of participants' perceptions. Newman's dimensions of defensible space relied on concepts drawn from across spatial and social fields. His book stated that his ambition was to draw together and incorporate material into an inter-disciplinary perspective:

> We have chosen to direct this work at a rather wide readership. It was initially intended primarily for housing developers, architects, city planners and police. But as the scope of the work grew and the significance of our findings became more apparent, it was felt that the manuscript should be reworked so as to make it more universally available. (Newman 1972: xiii)

His references were grouped into sections by discipline: environmental form, social policy, human territoriality, urban crime, housing and the sociology of the family. Reviews of *Defensible Space* showed that his concept provoked a strong response from an array of professions. Although each discipline/sector commended the overall idea, they tended to criticise it, perhaps not surprisingly, by emphasising the importance of, and seeking to protect/project, their own disciplines. The urban sociologist Mark Baldassare (1975) criticised Newman's research as 'the kind of sociologizing that is being done by other disciplines' (p. 435). Baldassare (1975) considered *Defensible Space* a methodologically sound study, however, he found Newman's interpretation of his findings 'sociologically naïve, or at least unproven' (Baldassare 1975: 435); particularly Newman's assertions that a collective identity would emerge to take responsibility for residential spaces. Finding Newman's architectural evaluation shallow, the urbanist Rayner Banham (1973) favoured the criminological over the urban design analysis: 'the non-architectural part is probably the more fruitful and meaningful of the two' (Banham 1973: 155). Planner John Friedmann (1973) emphasised the spatial limitations of Newman's analysis, arguing the publication was more about people's behaviour than spatial design. Friedmann's review unjustifiably asserted society's inherent impulses to crime and violence, and taking an unenthusiastic attitude to evidence gathering, asked: 'Why do we need costly scientific studies to prove to us what should be self-evident?' (Friedmann 1973: 49). Geographers and other planners said that defensible space was more likely to displace crime rather than eliminate it, questioning whether it could provide sustained and long-lasting improvements (Ley 1974a; Schneider and Kitchen 2002). Responding to this, the main governmental proponents of defensible space in the United States, HUD, argued that shifting crime away from areas of particular vulnerability made policing an easier task and that waves of defensible space strategies should be applied across a city, neighbourhood by neighbourhood (Cisneros 1996).

From the beginning, defensible space as a concept was disparaged for its negative framing of architectural form by the design professions, its views of human interaction by sociologists, its insensitivity to neighbourhood level social forces by urban geographers, and for its crude reading of crime by criminologists, such as concentrating on the analysis of offence rates independently of offender rates (Bottoms 1974). Nonetheless even Newman's most vocal critics agreed he was addressing 'a serious social problem' (Ley 1974a: 157), but that his crude interpretation ran 'a serious risk of debasing the importance of that theme' (Bottoms 1974: 206). Despite Ley's concerns that the significance of the idea was undermined by the poor quality of Newman's research, his review of Newman's book in the *Annals of the Association of American Geographers* recommended the exploration of defensible space as 'an intriguing and socially responsible task for the geographer' (Ley 1974a: 158)!

The term defensible space is now commonly used by a variety of disciplines. It has been subject to much cross-disciplinary scrutiny, not only from criminologists (Cozens et al. 2005; Reynalds and Elffers 2009; Taylor 1973), but also urban sociologists (Halpern 1995; Sampson et al. 1997; Wilson 1978a, 1978b, 1980, 1981a, 1981b), geographers of crime/criminal cartographers (Davidson 1981; Herbert 1982); social geographers exploring the geography of crime (Smith 1986a, 1986b); and urban designers/architects (Hillier 1973, 1986a, 1986b; Poyner 1983). Yet even within the broad topic of design and crime, Ekblom (2011) is careful to separate defensible space as a subset of crime prevention, spawning its own sub-fields such as environmental criminology. He defines defence of a space, building, object or person as both a 'preparatory' and an 'operational' task, with the role of defence and defender as further subsets. For a space to be defensible, it requires suitable physical properties, but defence equally has social dimensions, depending on the motivations, behaviour and capacity of the defender and offender. As this careful taxonomy suggests, criminologists' dissection of the mechanisms of defending spaces is extensive and thorough. Cozens et al. (2001a, 2001b) identified a hierarchy of four sequential levels: defensible space, undefended space, offensible space and indefensible space. A defensible space encourages a strong ownership response, yet undefended space (Merry 1981) may have similar characteristics (well surveyed, overlooked and inhabited) but for some reason is not actively defended. Offensible space (Atlas 1991) is still defended, but by 'others' in an anti-social way to facilitate criminal gang activity or drug dealing. Finally, indefensible space (Cozens et al. 2001a, 2001b) is where all social control mechanisms have broken down to such an extent that any design features are ineffective.

An alternative, and well-established, extension of the concept was into Crime Prevention Through Environmental Design (CPTED).[6] Ekblom (2011) traces the genealogy of the CPTED schools (e.g. Jacobs 1961; Jeffrey 1971; Newman 1972; Wilson and Kelling 1982) seeing its evolution as an accretion of ideas rather than progress towards a synthesised whole.

CPTED has seven components: territoriality, surveillance, image and management/maintenance, defensible space (as a distinct component from the former three), as well as target hardening, access control and activity support. These core concepts overlap, for example, territory is defended by controlling access to it, or surveillance influences opportunities for defence. These and other properties of the space (e.g. legibility, enclosure, porosity of boundaries) interconnect with design features to produce good or bad outcomes. So an assemblage of an isolated alley in combination with an overlooking window and a well-placed streetlight can mitigate criminal behaviour. Despite lacking clear evidence of causality, CPTED is attractively self-evident, with several decades of research finding design factors do work – just that these are *less* effective than other factors – recognising there may be other benefits from design interventions.

The concept of defensible space became quickly embedded in geography, for example, geographer David Herbert's (1982: 46) tripartite definition of defensible space, which summarised the range of its effects:

> as a model for residential environments which inhibit crime by creating the physical expression of a social fabric that defends itself; as a surrogate term for the range of mechanisms – real and symbolic barriers, strongly defined areas of influence and improved opportunities for surveillance – that combine to bring an environment under the control of its residents; and as a living environment which can be employed by its inhabitants for the enhancement of their own lives, while providing security for their neighbours, families and friends.

In fact, Herbert's definition neatly summarizes the dichotomy of disciplinary positions on defensible space: that of criminology (with its prime objective of inhibiting crime), and of planning/urban design (which aligns itself with the final, more positive, quality-of-life enhancing aims).

Geographers' interpretations of defensible space combined cartographical criminology (the spatial distribution of crime) with concepts from social psychology, particularly fear of crime. In *The Geography of Crime*, Herbert (1982) combined these by attempting to classify vulnerable urban environments and whether their occupants could identify their weaknesses. He found that particular design features of a dwelling did increase the likelihood of burglary, yet neighbourhood crime data revealed stronger patterns of social forces or inequality. Herbert's careful critique of defensible space concluded that the uncertainty around how it worked mattered little: 'if applications of defensible space ideas and associated social policies can improve the quality of life in city neighbourhoods and increase feelings of wellbeing, then these are in themselves ample justifications for such policies' (Herbert 1982: 110).

Various definitions of defensible space evolved from different disciplines – already within a critical context where various elements of defensible space were accepted/rejected depending on who the reader was. Yet despite the uptake of defensible space by different disciplines, the validity of defensible space as a concept was repeatedly questioned (Cozens et al. 2005; Ekblom 2011; Hillier 1973; Hillier and Sahbaz 2007). And, rather than completely discrediting it, this continuous critical inquiry has resulted in elaborations of an initially simple concept, in an attempt to explain better the perceptible effects. It is possible to trace back through this divergent history to a handful of foundational ideas (ownership, surveillance, interaction, territoriality). A striking aspect of this situation has been the malleability of the concept of defensible space; it has been promoted as a universal 'snake oil', before being attacked, refuted and disproved (interview with David Riley,[7] 2011). Yet, it remains a 'common-sense' concept, continuing to be used by architects and housing managers, and investigated by researchers with just enough agreement on what it constitutes for the idea to be applied in a workable way. The term 'defensible space' appears in newspapers, it is popularised as a positive recognisable housing attribute (*The Economist* 2006; Jenkins 2010). As with the professions, a basic version of defensible space is also recognised by the general public. It is this slippery ambiguity of defensible space that is intriguing, as is the way its mutability has helped this concept adapt as it has moved across disciplines, networks and policy domains. This is not to conclude, that defensible space is inherently unstable or endlessly flexible, more, as Larner and Le Heron (2002) describe, that concepts interacting in the world 'stabilise (become rationalities, meta-discourses, logics) as they are communicated and are instituted as the basis of action' (p. 720). Defensible space is a telling example of a concept that is unstable but highly durable and resilient as the basis of certain practical actions, yet less successful in retaining its logic when removed from this 'real world' by being transferred into formal policy.

Defensible space can be viewed either from Schön's (1983) theoretical high ground or from within his less rigorous 'messy swamps'; this is less a question of varied disciplinary perspectives than an indication of its mobile, mutable, uncertain nature. Cross-disciplinary interest in the topic has contributed to its definitional looseness but a greater definitional expansion can be traced from its use in differing regimes of practice. Practice is often seen as the messy, woolly obverse of academic theory. The simplified dualisms of theory in academia are helpful for polemic thinking and writing but they 'inhibit understanding by implying a certain neatness that is rarely found in lived life' (Flyvbjerg 2001: 49). In this book we reflect on the strengths, weaknesses and often hidden assumptions, inherent in the more familiar scrutinies of defensible space by disciplines such as criminology, planning or architectural design, reflecting on how theory and research shape policy and practice and vice versa.

Policy Mobilities

Unlike most of the policy mobilities literature (e.g. McCann 2011a, 2011b; Peck 2002; McCann 2013) that investigates pre-formed policies, we investigate the movement/mobility/mobilisation of a contested idea/concept. Indeed, 'defensible space' is a geographical concept – defensible *space*, not a policy per se. We look at how certain ideas about defensible space and associated strategies and views won through in the face of sometimes quite vicious criticism from academic peers. Critically we explore the daily mundaneness of policy formation, so often dominated by the use and misuse of evidence to justify decisions; this is unusual but necessary, as policy making is often embedded in the 'banal practices of bureaucrats' (McCann 2011a). Extending the policy mobilities work of McCann and Temenos (2015) on the practices and politics of public health we place equal importance on the mobility geographies of defensible space in English housing policy *and* practice.

We look at defensible space principles *practised* on housing estates in England, including a detailed case study of the Mozart Estate in Westminster, London (see Figures M.1–M.6). Here we draw on experiential knowledge as pragmatic rather than theoretical or scientific, orientated to here-and-now action. We also examine how defensible space consolidated political institutional settlement (Peck 2012) with respect to Thatcherism, Right to Buy etc., and update this to contemporary policies and practices, finishing the book with a discussion of the current English government's estate renewal programme (what Lees 2014, following Hyra 2008, has called the 'new' urban renewal) where defensible space still features. We explore the interaction of policy, the English housing sector and the state, in relation to the concept of defensible space.

Baker and Walker (2019) provide an excellent history of the multidisciplinary roots of thinking on policy movement, starting from what was known as 'policy diffusion studies' in the 1960s in the United States, which looked at intranational policy adoption (see Simmons et al. 2006; Simmons and Elkins 2004). Diffusion modellers were soon criticised for their focus on diffusion as resulting from processes of technocratic modernisation and competition, with policies carried along by rational, optimising agents or policy makers 'scanning the "market" for potential policy products' (Peck 2011a: 776). A less-blinkered interest in 'policy learning' then emerged in the 1990s from political scientists concerned that 'we know almost nothing about the process by which such policy transfer occurs' (Wolman 1992: 29). R. Rose (1991, 1993), for example, emphasised what he called 'lesson-drawing': the process by which policy makers learn from different sources. He was interested in the agents and actions associated with policy circulation. Rose (1991: 9) saw policy making as a process whereby policy makers set out to solve problems by seeking knowledge, and abstracting useful information from it; yet critically that they failed to achieve truly rational decision-making due to operating within institutional and political constraints. As Baker and Walker (2019: 5) state: 'with the turn to process-tracing, they [policy makers] began to open up analyses of policy circulation to issues of political contestation and strategic selectivity in policy making'. Subsequently, and also from within political science, 'policy transfer' emerged as a field of studies led by Dolowitz and Marsh (1996) who saw it as 'the process by which knowledge about policies, administrative arrangements, institutions and ideas in one political system (past or present) is used in the development of policies, administrative arrangements, institutions and ideas in another political system' (Dolowitz and Marsh 2000: 5). They were especially critical of the improbability of purely voluntary and direct policy transfers, arguing that the reality was much more messy. Although Stone (1999) had long argued that 'the process of modification in transfer requires closer attention' (p. 57), it took a while for this call to be heeded. But there was a gradual move away from policy transfer, towards work focused on how the 'form and effects' of certain policies are 'transformed by their journeys' (Peck 2011a: 793). From this emerged the notion of 'policy mobility', now known as 'policy mobilities' (however these terms are not fully interchangeable, see Soaita (2018) whose review shows that 'policy transfer' still dominates 'policy mobilities' terminologically in housing). This literature developed outside political science as a discipline, indeed was quite critical of it, taking a social constructivist approach, as we have in this book.

Soaita et al. (2021: 8) argue that policy mobilities scholars are 'more likely to look for thick-descriptions rather than explanations, aiming to unravel how policy is continually "assembled, disassembled and reassembled according to the elements (both discursive and material) that it encounters" (Lancione et al. 2017: 7)'. Policy mobilities thinking has been led by geographers, but also taken up by urban studies scholars, anthropologists and sociologists. Geographers are becoming increasingly aware of the social construction of policies, where and how they have been conceived, whether and how they have been mobilised and the political consequences of this (McCann 2013). As Baker and Walker (2019) point out, the organising metaphors of diffusion, lesson-drawing and transfer were replaced in policy mobilities by those of assemblage, mutation and mobilities (see McCann 2011a, 2011b; Peck 2011a; Peck and Theodore 2010a, 2010b). Those following a policy mobilities approach see policy as situated at any moment in time and as constantly contested. McFarlane (2011) talks about learning as 'situated seeing', where an individual's policy consciousness emerges via engagements with particular materials, people and environments. Geographers in particular have been interested in the spaces and spatialities in which policy makers do their learning (e.g. McCann 2011a, 2011b). By focusing on process policy mobilities scholars are exploring spatial multiplicity and dynamism, tracking how policy making processes 'have promiscuously spilled over jurisdictional boundaries, both "horizontally" (between national and local political entities) and "vertically" (between hierarchically scaled institutions and domains)' (Peck 2011a: 773). In the conclusion to their review of ideas on policy movement Baker and Walker (2019) argue for an interconnected trans-disciplinary future for what they call 'policy circulation studies': 'an agnostic, umbrella term' they use 'to denote research that examines the movement of policy knowledge and practices'. 'Policy circulation', for them, represents the state of play at the moment where a 'fragmenting pluralism' falls short of proper dialogic interaction across different research traditions and disciplines (Dolowitz and Marsh 2012; McCann 2011a, 2011b) and has become an obstacle to advancing knowledge on the what, how and why of policy circulation (Cook 2015; Dussauge-Laguna 2012). Jacobs and Lees (2013) have likewise been critical of the distinctions between policy transfer and policy mobilities: 'in making strident claims for the novelty of contemporary "mobilities" as opposed to past "transfers" assumptions are often made about how and why those earlier transfers happened and the effects they had' (p. 1561). The way forward, Baker and Walker (2019) argue, is through a commitment to an 'engaged pluralism', a commitment to intellectual openness, the creation of venues for dialogue and the (de/re)construction of coordinating concepts in policy circulation. In this book we have tried to do just that.

Like Baker and Walker (2019) we are also exploring disciplinary location in this book, but less in terms of how policy circulation has been seen by different disciplines such as political science, geography, anthropology or other fields. Rather our focus is on the policy or in our case the concept of defensible space itself, and how it was 'disciplined' in different disciplines (especially those with a tendency to intervention), such as criminology, geography and architecture. We are interested in how defensible space has been implanted, inserted, perhaps like a cuckoo in a nest; how defensible space has insinuated itself, like a Trojan horse, across different disciplinary frameworks.

We see the debates over defensible space from different disciplines as part of the process through which defensible space was mobilised. As such in engaging in the practice of engaged pluralism, we put on the table the multiple perspectives and critiques of defensible space as seen through the eyes of different disciplines. Later chapters describe how defensible space was mobilised differently in geography, criminology and architecture/planning. We answer Cook's (2015) call to listen to non-geography literatures (he mentions criminology and planning) on policy transfer and mobilities, but we go much further than that by looking at how a policy idea/concept was *mobilised* differently in different disciplines. Newburn et al. (2018) think through the value of notions of mobilities and assemblages in criminological debates; we do likewise providing a deeper and more extended historical narrative.

Baker and Walker (2019) argue that 'policy circulation' has three coordinating concepts: the arenas of policy circulation, the agents (these are necessarily diverse and include those who champion ideas, experts, global or more benevolent organisations, governments and civil servants, commercial bodies/economic actors etc.) and actions (including decision-making). In this book we focus on all three of these in some detail. But we also set out to think through three interrelated ideas: movement/mobility/mobilisation. Movement is the act or process of moving that can include change and development; mobility is the ability or not to move, the quality or state of being mobile; and mobilisation is the action of making something movable, mobile or capable of movement. The latter is about marshalling, organising, making ready for use or action. All of these three explain certain aspects of the story of defensible space. So far we have described the constituent pieces of the concept. As later chapters will show, it is the potential for alteration of these pieces during movement, the mobile state facilitating readjustment, that organises and finally assembles the various interconnected aspects of defensible space for suitable use in a specific context. In fact, mutability appears to be an essential defining characteristic of the concept. Yet this inherent instability or flexibility is not automatically negative. It is this adaptability and looseness of interpretation that has ensured its resilience in practice.

In order for policies to successfully circulate they must *be made* mobile (Peck 2011a); in contrast with the orthodox literature on policy transfer, 'the governing metaphors in critical policy studies are not those of transit and transaction, but of mobility and mutation' (Peck and Theodore 2010a: 170). Following Temenos and Baker (2015) call for more attention to embodiment, or 'a peopling of the geographies of policy mobilities', the book traces the movement/mobility/mobilisation of defensible space through certain people, namely a primary transfer agent: the geographer Alice Coleman; and, an attendant transfer agent: the Home Office (HO) researcher Sheena Wilson. As Peck and Theodore (2010a, 2010b) argue, policy transformations are clearly not realised declaratively or through administrative or authoritative decree or sanction – they are also embodied practices. But these actors are of course also embedded in specific contexts and in networks of association that channel learning potential (Peck 2011a, 2011b). Like Coleman, Wilson also met with Oscar Newman and given she worked in the HO this paved the way for his ideas to permeate there. Despite never meeting in person, Wilson's influence on Coleman is visible in the way she is quoted by name in *Utopia on Trial*. Wilson was a psychologist who had studied criminology and through her we see the transdisciplinary importance of defensible space, it is not just about architects' or geographers' takes on space and housing. Both of these transfer agents are coincidentally women. But considering Sheena Wilson, who worked in the HO and then the Department of the Environment (DoE), and her work with Michael Burbidge (senior principal research officer for the

DoE from 1975 until the late 1990s, when he supervised many influential cross departmental housing and crime studies), brings us to another female protagonist – Anne Power – whose early work stemmed from similar concerns about unpopular 'difficult to let' housing estates. But Anne Power, though acknowledging the role of design in successful housing, is a less direct transfer agent for defensible space; her influence on its dissemination is tangential in that more often she appears as a foil, especially to Alice Coleman.

Alice Coleman, Sheena Wilson and Anne Power all have practical influence and a desire to influence policy and guidance, but defensible space did not just transfer via powerful agents – it was also an obvious (common sense) idea that architects and planners picked up without knowing about its origins. In focusing on the key people involved in the mobilisation of defensible space into English housing policy and practice, we do not rely on Peck and Theodore's (2010a, 2010b) familiar idea of persuasive gurus (although this fits Oscar Newman well), indeed in Alice Coleman's case she could be considered as an 'anti-guru' (whilst acknowledged as a knowledgeable proponent, Coleman's forthright views often undermined colleagues acceptance of her in this position/role). Instead, in contrast to Peck and Theodore (2010a, 2010b) we look at the promotion and mobilisation of defensible space through individuals with particular interests in mind and succeeding despite resistance or opposition. Although not primary transfer agents, Sheena Wilson and Anne Power are not simply 'peripheral actors' (see Rusu and Löblová 2019); rather they are essential conduits providing the means for successful movement/mobility/ mobilisation. We see through all three (Coleman, Wilson and Power) how knowledge and expertise shapes careers and how individuals use their careers (both strategically and accidentally) to connect with new ideas, sites and contacts (cr. Craggs and Neate 2017).

Given that our primary transfer agent Alice Coleman is established as a counterfactual to contemporary positions and thought (an anti-guru), a well-connected 'outsider' working within the establishment, a degree of consideration of personality is necessary, particularly as one of our critiques of policy transfer thinking to date is the overlooked force of personality, even if the roles of charismatic individuals as transfer agents are acknowledged in the literature (McCann 2011a, 2011b). A brief comparison of Newman and Coleman focuses on their ability to influence and the tools they used to persuade, as much as their personalities themselves. Newman as salesman, applying architects' wiles, chatting with residents on film or using the advertising artifice of line-drawn images. Coleman's belief that the facts speak for themselves cultivated and reinforced by her dismissive brusqueness and reliance on indisputable interpretation of graphs, data and mapping. Like McFarlane (2011) we look at the socio-material assemblages that facilitated and inhibited the mobilisation of defensible space in policy and practice, from diagrams to graphs to letters to field trips to policy briefings, and so on.

The unusual fact that the primary and attendant transfer agents were all female urges us to employ a critical feminist perspective. Here we follow on from Gillian Rose's (1993) work on the nature of geographical thought and power–knowledge relationships. But this goes beyond simply revalorising the female protagonists, more fundamentally defensible space as a concept effected highly gendered responses to crime and fear of crime. Human geographers exploring the geography of crime have compared David Herbert's (1982) spatial mechanism led view of crime to Susan Smith's (1986b) feminist perspective that extended the social/spatial view of crime. More recent thinking on the geography of the fear of crime describes a subtle and complex sequence of interactions that link the design of the physical environment to perceived levels of violence, to fear of crime, and hence to an individuals' sense of general wellbeing. The subtlety of this interaction is open to individuals' interpretation not limited to design determinism (Smith 1986b, 2003). Nor does the book follow the typical feminist perspective exploring erasure due to difference. This perspective is a valid one, as Defensible Space and DICE is the story of many unnamed female housing estate residents whose lives were affected by the practical application of defensible space theories. Nonetheless some women's actions – such as Muriel Agnew of the Mozart Estate tenants' association – are recognised as championing sensitive approaches to regeneration (see the case study following Chapter 4). Here class and to a lesser extent racial/ethnic politics are important. Rather we explore a more subtle form of exclusion focusing on the delegitimization of ideas related to power relationships. Coleman's extreme and occasionally suspect ideas were not countered through persuasive logic or consistent analysis but via exclusion, derision and ostracism, which were in-and-of themselves attempts at erasure.

Unlike much of the policy mobilities research that has tended to focus on fast policies this book focuses on the slower and longer infiltration of a concept into policy and practice. Speed of mobilisation is important. Indeed, the notion of 'policy mobility' as different to its supposed predecessor 'policy transfer' implies speed, neoliberal speed at that (Peck and Theodore 2010b; see also Clarke 2012). The fact that much of the policy mobilities literature has looked at the rolling out of more recent neoliberal policies, such as workfare initiatives (Dolowitz 1998; Peck and Theodore 2010a), business improvement districts (Ward 2006; 2007; see also Cook 2008; Hoyt 2006; Tait and Jensen 2007), the creative city (González, 2011; Kong et al. 2006; Lees et al. 2016; Luckman et al. 2009; Peck 2005; Wang 2004) and healthcare programmes (McCann 2008; Ward 2006, 2007) is perhaps not surprising. But here we begin our analysis in a period when the neoliberal, privatisation agenda was just emerging (cr. Clarke 2012; Larner and Laurie 2010). Ours is a 'distended case' (Peck and Theodore 2012: 25) that follows defensible space across time and space and through localised instances. This stretching is sensitive both to relational and territorial geographies, geographies of flow and fixity and place specificities (Ward 2010).

In this chapter we have begun by tracing defensible space, if briefly, before it arrived in the United Kingdom. Much more work could be done on its origins (see particularly Knoblauch 2014, on the background economy of fear that influenced its success in the United States), but that is not the remit of this book. Rather, following Robinson (2015), our study is more interested in how it 'arrived at' various locations (policies, practices, even housing estates) via different people in England. We look at the multiple circulating stories about defensible space and the techniques used to create it from the notion arriving in the United Kingdom via Oscar Newman. But methodologically, like Gulson et al. (2017), we note the limitations of 'following the policy', because 'defensible space' was/is not a singular, cohesive concept, indeed it was/is not a policy. We trace how the concept was operationalised differently by a diverse set of agents (cr. Baker and Walker 2019), different academics, policy makers, practitioners and even industry.

Importantly, this is not a 'presentist' case study (many policy mobility studies are of contemporary policies), rather it is a historical narrative of an older concept and how it has been mobilised and moved over time. Here we echo Maddrell's call for 'historical work that is inclusive, contextual and theoretically and analytically sensitive to difference' (2009: 338). We focus in particular on the recent past – the late 1970s and the 1980s – which marked 'a key moment in the unfolding of the global privatization agenda' (Larner and Laurie 2010: 218); in so doing our study contributes to understanding the diverse geographies and histories of the privatisation of space through the global movement of theories, policies and techniques (Larner and Laurie 2010: 218; see also McCann 2008; Ward 2006). As such, it plays a modest role in extending our understanding of what Brenner and Theodore (2002: 349) refer to as the relationship between city building and 'actually existing neoliberalism'. Although the bulk of the book focuses empirically on the late 1970s and the 1980s, it also considers subsequent decades, taking the story of defensible space in English housing policy and practice right up to the current day, showing that defensible space continues to be mobile, mobilised and emergent.

Here we do not necessarily see the mobility of defensible space into English housing policy and practice as completely positive, indeed the use of this non-proven concept to argue for the demolition of 'sink estates' today is deeply problematic. Yet we do not see the mainstreaming of defensible space as entirely negative either. Its acceptance as a common-sense concept is powerful and inclusive. The problem is not only its uncritical use to justify estate renewal, but the unthinking application of urban design principles that can either lead to isolated exclusion or a more accessible, equitable and well-maintained public realm.[8]

Finally, we investigate what Allen and Imrie (2010) have called 'the knowledge business' in urban and housing research, in that we investigate the varied transdisciplinary nature of evidence use with respect to the concept of defensible space by policy makers and housing professionals. Similarly, by following Jacobs and Manzi (2013) and tracing the persistence of evidence-based policy's ubiquitous presence across administrations, particularly within English housing policy, we reconnect the often pragmatic practicalities of policy implementation with the highly politicised and uncompromising theorisation of what is perceived as good data and solid evidence.

The various bodies of knowledge we engage with (geographical, housing, architecture, design, planning etc.) all have shortcomings on how the mobility of policies might differ from the movement of concepts into research or into practice. We are interested in how ideas around defensible space developed through *doing* research, *making* policy and *practising* defensible space design on housing estates. In tracing the movement of defensible space from theory/concept to evidence-based practice, we ask a critical question: what does this mobility of defensible space tell us about how different communities of practice use evidence? As a result, we open up new avenues of inquiry for future policy mobilities work, taking 'a mobilities perspective [that] can extrapolate the specific ways in which everyday acts and technical practices are political actions that are ongoing and incremental' (McCann 2013: 593).

Researching and Writing about Defensible Space

few [*in policy mobilities scholarship*] are explicit about their methodological practice. (Baker and McGuirk 2017: 425)

It remains unusual in geography, even in the twenty-first century for two women to come together to write a book, especially in urban geography, which remains a male-dominated space. In fact, female voices permeate this book, not only our own but those of Alice Coleman, Sheena Wilson, Anne Power and many others. As such we see this book as contributing to the ongoing project of de-gendering geography (cr. G. Rose 1993), particularly urban geography, plus urban studies more widely. The more we researched the key transfer agents in this book the more we reflected on our own positions as women in the academy and in practice. We are both women with children (all daughters) and our writing by necessity had to fit around our family responsibilities and commitments (social reproduction), plus in Elanor's case a non-academic full-time job. Our researching and writing have been a pleasurable experience, collaborative, fun, critical and memorable. Although very interested in geographical and other theory we do not believe in theory for theory's sake, nor theory as a measure of intellect; rather we believe in grounded, empirical research and this takes a big commitment of time and energy.

As stated this is a deliberately trans-disciplinary book, as defensible space, although inherently geographical, also entered the disciplines of criminology, architecture and urban planning. Following the suggestion of Baker and Walker (2019) ours is a collegiate project of 'engaged pluralism', as we both come from different disciplinary backgrounds: Loretta is an urban geographer with over 25 years of research expertise in urban regeneration and architecture, Elanor is an architect who was Head of Research at the Commission for Architecture and the Built Environment (CABE) and who has also worked for housing associations including the Peabody Trust (where we first met). Loretta has been researching the regeneration of council estates for the past decade, working with residents to ensure socially sustainable regeneration; and Elanor has been working to quantify the long-term impact of regeneration projects on residents (including tools to evidence social value impacts), endeavouring to ensure that residents' wellbeing remains at the core of organisational financial decision-making. The book is a culmination of many years of research by both of us, indeed it draws on research undertaken in a number of different research grants over the years,[9] it also draws on Loretta's work in three public inquiries into the demolition of high-rise council estates in London, and Elanor's day-to-day research activities for the Greater London Authority (GLA), Clarion Housing Group and the wider G15 group of housing associations. This is an important and somewhat idiosyncratic historical story that has not been told in full before, there have been only fragments of the story in different places, here we bring those fragments together, plus much more, and hopefully the whole is more than the sum of its parts.

In this book, for the first time, we collate the disparate evidence on the mobilisation of defensible space in the English context. Our sources have been voluminous and in the end stages of writing this book we were given, very generously by the criminologist Simon Harding, yet more, new, material – a large body of research and material from the 1980s and 1990s on high-rise housing in the United Kingdom. Now this book is complete we are working on archiving this research and material and deciding what to do with it. Given the historical nature of this project we inevitably consulted a wide range of documentary materials, for as policy theorist Richard Freeman (2012) says, accessing policy mobilities requires a focus on constitutive practices of communicative interaction, both oral (in meetings) and textual (in documents), which he places as central to policy *making*, its production and reproduction. Our textual documents included:

- Media articles, both from the architectural and mainstream press, including the press cuttings of Coleman's publishers dating from the 1980s to the present.
- Policy documents dating from the 1970s that set the context for DICE, as well as more recent policy documents, to be able to trace the emergence of Coleman's ideas in subsequent planning, housing design and management guidance.
- The DoE's internal DICE reports.
- Market researchers MORI's numerical survey data for the DICE estates and their preliminary reports to Price Waterhouse for the DICE project.

- AEDAS Architects' project documents for the regeneration of the Mozart Estate.
- A member of the King's College London (KCL) DICE Unit's private archive of papers and reports.
- Alice Coleman's personal archive of all her research.

The first two categories on this list – publicly available media articles and policy documents were easier to locate along a critical spectrum as they tended to accentuate the particular attitudes of the publishers, or the audience they were addressing. Private papers, such as those of the KCL DICE researcher, suffered from a common drawback of informal archives in that they were highly selective and non-systematic, yet provided a candid interpretation of the situation. Although the central transfer agent – Alice Coleman's – work was quantitative, as was her DICE project and the large-scale quantitative and statistical evaluation that followed, and promoted as a highly scientific experimental process, our investigation has taken a qualitative approach. We undertook critical discourse analysis of the policy documents and attendant in-depth interviews with key academics and practitioners across the fields of architecture, criminology and planning, examining the varied viewpoints of these communities. Indeed, despite our argument being that defensible space is a concept lacking sufficient internal validity to be truly considered a theory, theory building for us was a critical frame for consideration. In the chapters in this book we trace how alternative physical and social ways of constructing defensible space and a kernel of shared principles have been reflected in research (both in Alice Coleman's DICE and Bill Hillier's Space Syntax), in housing policy (in the form of funding documents for the Estates Action Programme [EAP]) and finally codified into design guidance (the DICE Design Guides, DoE documentation and the early versions of SBD).

Although much of the research underpinning this book is historical – that is, recent history – many of the key protagonists in the story are still alive, for example, Alice Coleman (b.1923) is in her late 90s. As such there were, and remain, ethical considerations. Throughout Coleman's career she often forthrightly defended her opinions, which politically at least have long been counter to those in her home discipline of geography; hers has been seen as a maverick viewpoint. Coleman and her work were seen as politically embarrassing to the Department of Geography at KCL where she worked, and to human geography more widely. As such we need/ed to take care of Coleman, given her age and potential vulnerability to personal criticism, we need/ed to take extra care to treat her and her work sensitively and with as open a mind as possible. As Jacobs and Lees (2013) point out, Coleman was engaged with what many consider to be the wrong brand of 'public geography' (Burawoy 2004, 2005): someone operating in the 'extra-academic realm' conducting scholarship that was concrete, pragmatic and serving wider neoliberal policy agendas and clients (see also Castree 2006; Fuller 2008). That scholarship appeared increasingly out of step with the trends then restructuring ([critical] geographic) academic thought.

Maddrell's (2009) useful discussion of Coleman's geographical work finishes in 1970 when ours starts, and her impact on English housing policy has not been discussed in any detail in histories of geography or housing, remaining an interesting omission. Coleman was an unusual geographer – she trained initially as a physical geographer, became a human geographer[10] (we look at the skill sets she brought between the two) and was a Tory at a time in the 1980s when many (but not all) geographers were quite the opposite. We look at her networks in the discipline and how these made her the woman she was/is. She was also a woman and eventually a professor at a time when geography was still very male dominated and the cultural turn was only beginning. She was the first female professor of geography at KCL, but we hear from a different, non-feminist, right-wing voice on these issues, something that the discipline has not done to date. Soaita et al. (2021: 11) argue that 'policy moves "successfully" across places and actors of similar ideological sensitivities', but as our research shows – yes and no.

For the first time, our book critiques some of the vociferous critiques of Coleman herself; it highlights the fact that a female professor of geography won a research grant of £50 million (a substantial grant now and unprecedented 40 years ago), and that her ideas have had arguably the biggest lasting impact on housing policy and practice of any recent geographer. A review of Danny Dorling's book *All that is Solid: How the Great Housing Disaster Defines Our Times and What We Can Do About It* (2014), chides Dorling for overlooking Coleman's work (amongst other female urbanists), particularly Coleman's critique of pre-1970s council housing (see Domosh 1991, on such absences). The review says 'for Dorling to have ignored these feisty predecessors makes his work questionable' (Bar-Hillel 2014: 46). We do that work in this book. Coleman, ironically, was once awarded *The Veuve Clicquot Award* for being 'A woman in a man's world' yet she had little regard for feminist agendas: 'I was a geographer rather than just a female', she insists (interview with Alice Coleman, 2008). But including women in geographical knowledge and history (G. Rose 1993; Maddrell 2009) is of course an ongoing project! Here, our focus on Coleman does HERstory (see Barnes and Sheppard, 2019).

Coleman, however, is only one of many feisty and successful women in this story; Sheena Wilson is discussed as another key transfer agent, and Anne Power as Coleman's nemesis. Jacobs and Lees (2013) only discuss one transfer agent, Coleman; here we extend their study underlining the multiplicity and messiness of transfer. In 'exploring' Coleman in relation to other (female) actors in the story we also necessarily avoid concentrating any critique on her personally.

In the book we draw on a series of in-depth interviews with Coleman: in 2008 Loretta Lees and 'the other' Jane Jacobs, a geographer, recorded a half-day interview with Coleman in Loretta's then office at KCL, they also video recorded a half-day 'walk along interview' in which Alice was interviewed re-visiting some of the public-housing estates in East London where she put into action her version of Newman's idea of defensible space. In 2010 they also undertook a video and tape-recorded day-long interview of Coleman discussing her work in and through her own personal archives in her home in Dulwich, South London (see Figure 1.3). Coleman had an extensive personal archive of the research undertaken during her career, housed in a separate house, adjacent to her home. Then in 2013 Elanor Warwick undertook follow-up interviews with Coleman, firstly after her 90th birthday celebration at KCL, where she reflected on themes such as her drive and focus for research, her desire to be a teacher and her struggles to access crime data and persuade funders and officials, including Michael Heseltine, of the value of her work. These were indications of her single mindedness and persistence but lead to what she considered her proudest achievement – 'bringing the crime rate down with Thatcher'. These themes were explored in later interviews again in her home. Nicholas Boys Smith of Create Streets[11] in 2014, who like us, found:

> At 91 Coleman remains a formidable force. The fruits of a keenly inquiring mind and productive professional life are evident all around the room, ranging from a unique keyboard developed to write a phonic version of the alphabet that Coleman has developed herself, to stacks of large glossy colour plates of her exhaustive land utilisation survey. Only 15 percent was ever printed. The rest is in her dining room or (carefully stored under blue tarpaulin) outside on the terrace. It is rather alarming to think that the only copy of one of the most accurate surveys of the UK ever conducted is stored at partial risk to the elements in suburban London. Her memory is strikingly sharp and fast – and wide ranging.[12]

Figure 1.3   Loretta Lees' and Jane Jacobs' interview with Alice Coleman in her archives. (Photograph: Loretta Lees)

He finishes: 'Alice Coleman watches the world cheerfully and with remarkable precision from her Dulwich home but she wonders why this generation is repeating the mistakes of her own' (it is worth watching Alice in this interview https://vimeo.com/ondemand/designedforliving to get a sense of our protagonist and her views).

Of course, all interviews have inherent bias, and reflections on past events involve seamless post-facto rationalisations in which ambivalence, multiple motivations, dilemmas and failures are concealed (cr. Chamberlain and Leydesdorff 2004). Coleman was somewhat guarded with respect to the scepticism of peers and the DoE whom she sees as having hindered her vision for design intervention in council housing. As such we had to triangulate Coleman's account by looking beyond her recollections and archive to other sources, including interviews with national and local government figures involved with Coleman or dealing with the implementation of programmes of action based on her findings. We have also looked at media and scholarly reviews of her work, produced either as stand-alone reviews or as part of historical housing and public policy review accounts; articles in architectural or planning magazines, as well as mainstream press. This diversity of sources and literatures, spanning academic journals to red-top newspapers and TV programmes, threw up the challenge of reconciling contradictory presentations but flagged up the reach of the idea and the breadth of applied domains and practical vocations it touched. It was this multiplicity of contexts that made discourse analysis a suitable technique for exploring the disparity of viewpoints between academics, practitioners and politicians. In fact, Jacobs and Manzi (1996) suggest that the process of policy making itself can be understood as a form of discourse analysis. Discourse analysis then, was used both as a technique for paying close attention to the evolution and translation of concepts, and also as a conceptual bridge between practice and the mechanisms of transfer.

A key consideration when writing any book is who is it for? Earlier we quoted Oscar Newman who wrote his defensible space book for 'a diverse audience'. Of course, our core audience are geographers – those interested in urban geography, housing, urban regeneration, architectural geographies, the geography of crime, the history of geography, and so forth. But the book has not just been written for geographers, it is an inherently trans-disciplinary book, written for others in urban planning, architecture, criminology, sociology, environmental psychology, and so on. Its target audience is also built-environment professionals and housing practitioners, policy makers, those involved in environmental design and 'secure by design' consultancies, and urbanists more generally. Looking in some detail at 'defensible space' in practice, and evaluations of it, offers some useful insights for practitioners on its value or not. Finally, the research and findings in this book have real potential in terms of informing government funded programmes of work on the demolition and/or remodelling of urban/high-rise estates in the

United Kingdom, especially in reference to defensible space and thus helping those fighting this 'new' urban renewal.

The book tracks defensible space from being a somewhat simplistic, almost naive concept, as it goes through a gradual process of elaboration into the current, multifaceted, multi-layered version. Cross-disciplinary interest in the topic adds to this definitional expansion and suggests that reinterpretations will continue. The book begins properly, in this chapter, with Newman's definition of defensible space in New York City in the 1970s (recognising that its origins are much earlier) as it is about to enter English housing policy and practice. We then turn to its mobilisation in Chapter 2 through Newman himself, but also through two key agents – Sheena Wilson and Alice Coleman – who also meet with Newman, if separately. The book investigates the period from 1970 to 1990 when defensible space was examined by academics and designers and was at its peak of influencing policy and design guidance, despite being contested. Chapter 3 looks in detail at the push back against Coleman's reinterpretation of Newman's defensible space in the UK context, including Coleman's personal reflections on this negative response, but we also look at how her interpretation attracted those in power. In Chapter 4, we explore Coleman's DICE (the Design Improvement Control Experiment) project, the remodelling process dubbed 'Colemanisation' by the *Municipal Journal* (1990). In Chapter 5 we consider the evaluations of defensible space in the Priority Estates Project (PEP) and the EAP, the Price Waterhouse evaluation of Coleman's DICE and Coleman's own evaluation. The final chapters bring the investigation up to date, Chapter 6 looks at the uptake and resilience of defensible space ideas since DICE, from SBD to estate renewal from the late 1990s onwards. The latest, from a designing out crime standpoint, is Armitage and Ekblom's (2019) reformulation of CPTED, which has defensible space as a foundational element. This attempts to reinvigorate the somewhat 'one-size-fits' CPTED principles, by integrating situational crime-prevention, offender-related approaches with design, propelling 'CPTED towards a properly evidence-based, theory informed, conceptually sound and practically feasible field of intervention' (Armitage and Ekblom 2019: 3). We extend this ambition in Chapter 7, reflecting on the relationship between research, policy and practice, through a discussion of defensible space as a mid-range theory. We ask if the way in which defensible space was mobilised affected its impact and how it was, and is still, used to construct places to live. This reveals pertinent lessons on mobile concepts for academia and policy, highlighting the value of policy mobilities research. We conclude with examples of contemporary homes and their surrounding estates that demonstrate the negative outcomes of applying exaggerated dimensions of defensible space at increasingly high densities. By speculating where present practice may lead, we show this is not the end of the story of defensible space, which, irrepressibly, will no doubt continue to evolve.

Notes

1   *Estate regeneration: article by David Cameron.* https://www.gov.uk/government/speeches/estate-regeneration-article-by-david-cameron.

2   Our focus is on English housing policy and practice because there is no such thing as 'UK Housing Policy'. Answering Pinch (1998) we offer a more 'spatially aware' analysis that avoids conflating the English experience with that of the United Kingdom as a whole. Housing policy has of course been a devolved matter since 1999 and prior to the establishment of the devolved administrations in Northern Ireland, Scotland and Wales, separate laws and regulatory frameworks existed across the United Kingdom.

3   Newman's classifications were: interior spaces within flats are private and streets public. Lobbies, stairs or shared internal spaces are semi-private spaces and external gardens accessed by a number of residents, semi-public. However, depending on the design and layout, internal circulation areas could be classified as public. Coleman's later categories included 'confused' space, corresponding to the planning term SLOPE – space left over from planning.

4   Note this is not the normal reading of 'routine', to criminologists 'routine activity theory explains that a crime event occurs when a motivated offender and a suitable target converge in time and space in the absence of a capable guardian' (Cohen and Felson 1979).

5   *Defensible space and the open society.* http://www.we-aggregate.org/piece/defensible-space-and-the-open-society.

6   Jeffrey (1999) who originally devised the term 'crime prevention through environmental design' in 1971 stated that its principles were largely based on Newman's work.

7   David Riley was the civil servant managing the DICE research project at the DoE.

8   See The Urban Idiot (2015) Urban idiocy: Brilliant ideas that ruined our cities. Part two: Secured by Design, *Here and Now: Academy of Urbanism Journal*, 5 (Spring), p. 54.

9   Research from the following projects underpin this book: PI: Lees, L., CoIs: Hubbard, P. and Tate, N. ESRC 2017–2020. Gentrification, Displacement, and the Impacts of Council Estate Renewal in C21st London *[ES/N015053\1]*; PI: Lees, L. CoIs: London Tenants Federation, Just Space and Southwark Notes Archive Group, Antipode Activist Scholar Award 2012. 'Challenging "the New Urban Renewal": Fathering the Tools Necessary to Halt the Social Cleansing of Council Estates and Developing Community-led Alternatives for Sustaining Existing Communities'; Lees, L. (PI) 2003–2006 'High-rise Living in London: Towards an Urban Renaissance' from ESRC under the ODPM studentship programme; Lees, L. (PI) 2001–2004 'Young People, Place and Urban Regeneration: the Case of the King's Cross Ten Estates, London' from ESRC under CASE studentship programme, collaborating partner – the Peabody Trust; PI Warwick, E. KCL PhD 'Defensible Space as a Mobile Concept: the Role of Transfer Mechanisms and Evidence in Housing Research, Policy and Practice', funded by CABE; and finally, unfunded research undertaken with Professor Jane Jacobs (then at the University of Edinburgh) on Alice Coleman.

10 The Royal Geographical Society presented Coleman with the Gill Memorial Award (1963) and Busk Award (1987).

11 See Create Streets http://dev.createstreets.com.

12 Boys Smith and Wildblood (2014).

Chapter Two
Defensible Space Is Mobilised in England

The publication of Oscar Newman's *Defensible Space: Crime Prevention Through Urban Design* (1972) received a great deal of attention from the US press and television. It was extremely accessible, illustrated by artfully composed photographs and sketches, and Newman's appealing narrative style presented the idea in a comprehensible, inviting manner to an academic and lay audience. The US architectural profession reacted positively to the publication, linking it back to the utopian ideas of Lewis Mumford, Kevin Lynch and other architectural writers (Friedmann 1973). Defensible space 'had a good brand, had a good name, it had everything going for it' (interview Sheena Wilson, 2012) and the concept moved rapidly across the Atlantic. In fact, Newman's book was re/published a year later in England under a new title, *Defensible Space: People and Design in the Violent City* (1973) with a more theatrically dramatic cover (see Figure 2.1).

To promote publication of his book in England, Newman publicised his ideas via a UK book tour and lectures. One lecture, at the University of Sheffield, was attended by Rob Mawby, then Lecturer in Criminology, working then on the *Sheffield Study on Urban Social Structure and Crime*. During this trip, Newman visited several British housing estates analysing them according to his defensible space theory (Mawby 1977). Newman presented his findings from these visits at a conference held in London in December 1974: *Architecture, Planning and Urban Crime*. The conference was organised by the National Association for the Care and Resettlement of Offenders (NACRO) with the Royal Institute of British Architects (RIBA) and the Royal Town Planning Institute. Mawby reported that this

Defensible Space on the Move: Mobilisation in English Housing Policy and Practice, First Edition. Loretta Lees and Elanor Warwick.
© 2022 Royal Geographical Society (with the Institute of British Geographers). Published 2022 by John Wiley & Sons Ltd.

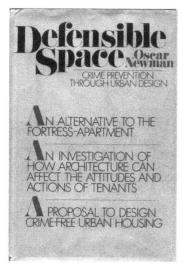

Defensible Space: Crime
Prevention Through Urban Design
(1972) New York: Macmillian

Defensible Space: People and
Design in the Violent City
(1973) London: Architectural
Press

Figure 2.1   *Defensible Space,* US and UK editions.

appearance prompted demands for a replication of Newman's New York study in England (Ash et al. 1975). Nevertheless, criminologist Mawby was dismissive of architects' and planners' understanding of crime, citing the DoE researcher Mike Burbidge's 'rather inadequate contribution to the NACRO conference compared for example with [criminologist] Baldwin's more detailed analysis' (Mawby 1977: 178). In fact, Mawby included Newman in this category of inexperienced architects/planners, condemning his crude research techniques and selective reporting: 'It is not sufficient to cite an advantageous factor and ignore a disadvantageous one' (Mawby 1977: 177). Mawby re-examined his Sheffield data in the light of Newman's theory and found little evidence to support the theory that high-rise flats were more prone to crime because of a lack of defensible space. However, he was more supportive of Newman's explanation for the increased tendency to report witnessed crimes, which echoed Jane Jacobs' (1961) 'eyes on the street' and appeared to be affected by how much residents believed they were living in a high crime area. Mawby complained that '"crime" as described in Newman's *Defensible Space* is invariably an emotive concept, graphically illustrated by such terms as "vandalism" or "mugging"' and called for a cooler more objective discussion of offenders, offence rates and possible causality (Mawby 1977: 175). Critically, Newman was seen to reduce 'the criminal' to a social stereotype, an inveterate and perpetual lawbreaker (Campkin 2013). The background context of urban crisis in England in the 1960s and 1970s and the emergence of the deindustrialising, decaying, inner city as 'a problem' was relevant. Indeed, it provided rhetoric for the law-and-order lobby and 'Oscar Newman was to provide everyone's salvation by designing "defensible space"' (Cross and Keith 1993: 203). Race was used to conjure up images of urban crisis (Gilroy 1987; Solomos 1988), with the tabloids using racist images of black criminality.

Despite the criticisms Newman's ideas spread extensively throughout England and in 1978 defensible space had 'become common currency amongst housing managers, architects and even tenants' (Wilson 1978a: 674). Outside academic circles Newman's book was reviewed in national and local newspapers. One article in a local Sheffield paper, titled *High Rise Flats are Nurseries of Crime* uncritically reported, without question, the reliability of the data but also that the findings were transferable to the British context (Brown 1974). The architectural press was far less favourable. The extensive US press machine behind the publication of the book had generated high expectations but the architectural reviewer in *Building Design* magazine found it a disappointing book providing few new insights. The review exuded a sense of British superiority: 'the intolerable situation of crime and violence in America is taken at face value ... The rather extreme problems of public housing in America' (*Building Design* 1973: 8), complaining that Newman expected architecture to solve 'problems of a political nature' (*Building Design* 1973: 8). Dismissive reviews were many, for example, in *Architects' Journal* architect, planner and anarchist Colin Ward (1973a) reproached Newman for 'propagating a crudely deterministic approach to urban design; architecture as a branch of police science', despite Newman calling for greater resident participation (Ward 1973a: 1243). These accusations of determinism were commonly repeated, for Newman's ideas were considered a 'novel and contentious brand of architectural determinism' (Wilson 1978a: 674).

This sense of discomfort can be attributed to the book exposing serious weaknesses in the built-environment profession's understanding of the impact of building design. Architect Bill Hillier,[1] writing in the *RIBA Journal* at the time argued that the book was:

> really about the crisis in our knowledge of the relationships between the forms of artificial space we create and the social behaviour that goes on in it – knowledge that architects take as their stock in trade in order to design anything. In other words *Defensible Space* is a bad book about a very important subject. (Hillier 1973: 543)

Some of Newman's critics saw his promotion of defensible space as a social 'movement' rather than a robustly applied theoretical framework: 'Newman's writing style is predisposed to political oratory rather than serious scientific endeavour' (Mawby 1977: 169). Reviews such as this exhibited a degree of professional elitism, considering Newman's presentation of the concept better suited to addressing a general, less-knowledgeable audience, rather than well-informed professionals.

The 1974 BBC *Horizon* program 'The Writing on the Wall' introduced New-man's ideas to a wider British public (Mansfield 1974). Newman, using the title Professor to establish credibility, was filmed either at his office desk, overlooking dramatic New York skylines, or photogenically striding around 'disreputable' housing projects in a sweeping long black coat. 'Newman's showmanship was notable' (interview Sheena Wilson, 2012; see Figure 2.2) and the programme had a theatrical quality. It opened with images of the poor conditions in New York housing projects (as well as of Pruitt-Igoe in St Louis, Chicago) comparing crime statistics in New York City to British ones, reinforcing perceptions that US crime problems were worse than in England. Pictures of a poorly maintained housing block followed, where it was 'catastrophically clear' that the physical design should be blamed; the ugly entrance ways, 'these cells called elevators' and an outside play area described as a 'cage for children'. But in a suitably tele-visual shock, the narrator announced 'This isn't New York – this is England'! The Aylesbury Estate in London was not actually named as the location, but could easily be identified from a section filmed of Newman walking through the streets surrounding it. He described the estate's recently completed, grey white mass as a 'creature from another world', its long horizontal concrete blocks towering above the neighbouring terraced houses and blocking off the surrounding streets. We will return to the Aylesbury Estate, one of the largest public-housing estates in Europe and under order of demolition (see Lees 2014), in later chapters where we discuss the current Conservative government's sink estate policies and demolition of council housing, especially high-rise council estates.

Interviews were undertaken in the *Horizon* programme with both unhappy US *and* British public-housing residents, who voiced similar complaints. Newman is also filmed chatting with a group of elderly ladies who explain how much they

Figure 2.2   Still of Oscar Newman on the 1974 BBC *Horizon* program 'The Writing on the Wall'. (Permission: Kopper Newman)

enjoy living on their estate, but he uses this as evidence that tower blocks can successfully house elderly residents without crime, recommending against mixing families and older populations. The programme's focus repeatedly returns to children: film clips of poorly designed, neglected play areas, children playing unsupervised in unsafe locations.[2] Newman's proposed solutions were a mix of careful allocation of residents, external design interventions to blocks/spaces and greater use of surveillance technology, particularly CCTV and intercoms. Throughout the programme Newman rarely mentions defensible space, but talks about the 'definition of space' and those 'who overlook the space'.

As his television appearances show, Newman consciously presented his principles in a persuasive, appealing manner. This deliberate salesmanship echoes other descriptions of transfer agents as policy brokers or entrepreneurs (see Peck and Theodore 2010a, 2010b; Rydin 2003):

> He was an unusual figure, quite charismatic. He wasn't like a social worker who was really committed to people, he just had this theory that he sold. He was more like a businessman selling the defensible space concept. He was successful. He was travelling abroad a lot. He was almost like a social entrepreneur' (interview with Sheena Wilson, 2012)

The *Horizon* programme struck an odd balance between supported evidence and unsupported assertion, with Newman disingenuously describing the children filmed as the first generation of British children to grow up in high-rise homes. As the first tower blocks were constructed in England immediately after the Second World War, there had been at least two decades of young inhabitants, potentially into a second generation. Tellingly, Newman attributes an inevitable progression from vandalism to more serious crime: 'There's no clear evidence in England on vandalism, but one wonders will these children grow up to be criminals?' (Newman in Mansfield 1974). This may have been dramatic emphasis for the programme's finale, but it was an understatement of the research understanding and level of policy interest in the issue of vandalism in England at that time. Newman's photographs of the housing projects' vandalised entrances, lifts and lobbies were a highly recognisable feature for British architects, planners and housing officers, who easily transposed them to their own experience of council housing. One chapter of Poyner's widely quoted book *Design Against Crime* (1983) discusses vandalism within public housing. Here he explores the British fixation on vandalism, complaining that despite the widespread attention, it was too often dismissed as a minor example of anti-social behaviour rather than a crime with serious impact. He forcefully reiterated the variation between the United States and Britain during the 1970s: in Britain more than a third of all housing was within the public sector, whereas less than 3% of North American housing was managed by the state, the majority of which was perceived to be 'problem Projects', subject to the most extreme forms of crime. Yet this difference was narrowing and both the incidence of, and fear of crime, was increasing at unprecedented rates (Poyner 1983). During the 1970s the positive legacy of Aneurin (Nye) Bevan's post-war public-sector housing programme (see Boughton 2018) still coloured views of council housing with only a minority being perceived as 'difficult-to-let' estates. Yet while British council estates had far lower actual crime rates than in the United States, the degree of public concern about vandalism was extremely high at the time, contra Newman.

There was extensive debate on the issue in the mainstream and the professional press amongst writers such as Laurie Taylor and Colin Ward (Downes 1974; McKean 1973; Taylor 1973; Ward 1973b). The research establishment cited the visible indications of damage to justify further investigation, even if the vandalism consisted of numerous trivial incidents. An illustrated Design Council guide *Designing Against Vandalism* (Sykes 1979) gathered together research summaries by then DoE researcher Sheena Wilson and others setting out practical lessons for Local Authorities and housing managers (Wilson 1979). Newman's lack of awareness of the emerging British evidence – the DoE research into vandalism (Burbidge 1973) and the evidence gathering beginning at the HO (Sturman and Wilson 1976) – can be read as a failed example of international transferability, with reputable English research not being disseminated in the United States or the inability to construct a shared knowledge base due to closed or disconnected epistemic networks.[3] The focus of epistemic networks on creating knowledge across a specific domain and establishing robust and mutually accepted claims can result in overly inward-looking groups (Cooper 2006; Rydin 2003). But to be fair to Newman, competitive barriers existed between academic and professional networks in the United States and England. Likewise, the government research teams, while exploring the same topics and participating in similar networks, did so with differing policy drivers and hence there was a tendency towards siloed working. There were reports at the time of sparring between the different government departments interested in inner cities, most seeking to extend or defend their spheres of influence. The sparring was aggravated by their different philosophical approaches: 'despite all the machinery for inter-departmental co-operation, departments have been showing a marked tendency to take a tunnel vision approach to the pursuit of their own urban policies' (Cameron 1987).

Preparing the Way for Defensible Space

Oscar Newman and Sheena Wilson were from very different research cultures and political backgrounds. They may have been studying similar housing typologies, but at the time the United States and England were very different housing and policy contexts. Nonetheless Sheena Wilson became an influential transfer agent for defensible space when she brought Newman's ideas to the attention of a wide policy audience, and played a critical role communicating ideas between the HO and the DoE. Wilson trained as a psychologist and studied criminology before joining the HO Research Unit in 1971 at a point when it was ceasing psychology-based studies and beginning to explore socio-economic motivations for crime; moving away from the deep-seated idea that there was a criminal 'type' to one where much crime was opportunistic. This alternative opportunistic model of crime opened up ideas of crime prevention, where police advice and good design practice could intervene directly in the mechanics of crime reduction (Mayhew et al. 1976). In parallel with the DoE, the HO began to take a growing interest in vandalism as an example of opportunistic crime. Wilson's early research set out to investigate the causes of vandalism on housing estates (Sturman and Wilson 1976). She notes that while the research was as quantitative as possible for the time, it was in fact rather primitive:

> We tried in a very scientific way to set up a sample of different types of buildings in different areas. We used regression analysis on one of these really big computers, but we couldn't control for the background of the people because it was too hard – for one there was no computerized information. I literally went through housing records kept on index cards in each housing department. (interview with Sheena Wilson, 2012)

The DoE's housing research section dated from the mid-1960s; during his period as Minister for Housing and Local Government, Richard Crossman, wrote in his diaries of walking around his departmental empire and coming across a 'centre of excellence', a small left-wing group of thinkers which was the Social Research Division (SRD), later under the direction of Judith Littlewood to become the Social Research Branch (Crossman 1975; interview with Toby Taper, 2013). The remit of SRD was more engaged with political policy formation than the group of architects within the Housing Development Directorate (HDD).[4] While both commissioned and undertook housing research, HDD was more responsible for devising technical housing guidelines and Building Bulletins, including the highly influential Design Bulletin 25 *The Estate Outside the Dwelling: Reactions of Residents to Aspects of Housing Layout* (Reynolds et al. 1972). This was the first of several bulletins dealing with the 'problems encountered' within public housing and was based on a 1967 quantitative survey of residents in six medium density local authority housing estates in London and Sheffield built since 1960. The study found that housewives' overall satisfaction with living on their estates shaped their views of their estate's appearance. This over-all satisfaction was a composite made up of satisfaction with the landlord (main-tenance regimes, response to repairs and ability to deal with social annoyances) combined with the layout of the estate and the spaciousness of green areas. The seven variables most strongly influencing satisfaction were: appearance of the estate, the dwelling, maintenance standards, opportunities to move, problem play areas, the view from the flat and vandalism. Additional significant variables related to the concept of defensible space included access to private space, too many people and blocks being too large. In fact, many of the principles that Newman (and later

Coleman) would raise as potential problem areas had been identified in this earlier research. Surprisingly, height above the ground was found to have no impact on housewives' satisfaction with their home. Space for children's play was highlighted as a significant area of concern, regardless of the quantity or design of play areas provided. The authors surmised this dissatisfaction was caused by maternal concern for children's safety and supervision. Yet the authors observed children playing alone throughout the estates and that most spaces, whether intended for children's activities or not, provided opportunities for play.

About a quarter of the residents surveyed were dissatisfied with maintenance and over half felt that vandalism was a problem. This was attributed to insufficient staff being available to supervise the estate, especially preventing children 'spoiling the place' and making it 'dirty, with litter all around' (Reynolds et al. 1972: 21). Vandalism, which was classified as damage to property as well as general 'untidiness', was perceived as a widespread problem. Complaints covered damage to lifts and telephone boxes, uprooting of trees and flowers, breaking milk bottles or windows and noisy teenagers, but graffiti was not a major concern. The discussion identified that damage was seen less as wilful destruction than the result of children's thoughtless behaviour, taking short cuts across flowerbeds or dropping litter. Interestingly more complaints were made about the estates with the smallest number of reported occurrences, suggesting that it was the contrast with the overall appearance of the estate being noticed. These views that vandalism was a social problem as much as a physical one contradicted Newman's opinion that poor design led inevitably to intentional damage!

In 1977, whilst working for the HO, Sheena Wilson travelled to the United States and spent a day with Oscar Newman looking at housing projects in the Bronx, accompanied by a police escort. Two points struck her: the extreme levels of poverty and vandalism occurring in parts of New York City compared to in England (post-war decline and New York City's bankruptcy in the 1970s triggered a deteriorating urban environment, particularly in public housing) and that the manner in which improvements were implemented affected the sustainable long-term success of the changes:

> These streets were really dangerous, it was like Armageddon, riddled with potholes, there were burnt out cars, fire hydrants flowing, I've never seen such social desolation …We were looking at a 3rd generation permanent underclass, unemployed young guys with no teeth and on drugs, sitting around in stairwells accosting people for money. Back then in the UK we were tackling something very different: the housing design of places like Hulme, industrial system building combined with limited housing management, social inadequacies and the sheer density of children.
> (interview with Sheena Wilson, 2012)

Wilson and Newman revisited Clason Point Gardens and Markham Gardens in New York City, two projects that featured in *Defensible Space*. The defensible space improvements to Clason Point Gardens had relied on engaging and working with existing residents. To test the impact of this community involvement, Newman carried out the alterations at Markham Gardens in a more traditional authoritarian way, with workers appearing unannounced and erecting fences. Residents did not respond well to the imposition of these unexpected new barriers and the new landscaping had suffered substantial damage as a consequence. Writing about the visit Wilson recalls 'the style with which physical improvements were implemented was more important than the measures themselves' (Wilson 1978a: 674).

Refurbishment schemes with a too literal application of his principles were found to continue to have vandalism problems. Local Authorities who had simplistically altered the design of estate layouts without improving their management services were disappointed at the lack of territoriality or pride of place their changes generated. Ironically, Wilson felt that it was the process of testing the concept of defensible space that led the DoE to a more robust belief in the importance of non-design factors:

> Ironically it was speculation inspired by defensible space which eventually led to a clearer understanding of the non-design factors contributing to the success of housing schemes ... The inconsistency of the evidence implied that design should never be considered independently of social and management factors. (Wilson 1978a: 674)

Wilson acted as a common, overlapping thread between different government departments in the mobilisation of defensible space. Despite her scepticism, Wilson had some enthusiasm for Newman's ideas, and Newman's initial concepts were mobilised (made ready to use/enacted in a different policy context) from merely reducing the physical settings and opportunity for crimes like burglary, to enhancing communities' ability to police themselves. Wilson's criticism of much contemporary criminological research into inner-city problems was the concentration on causes and blame, which imposed solutions in a top-down way rather than allowing diverse community-led solutions to emerge to tackle the multiple challenges: 'There is no such thing as *a single* solution' (interview with Sheena Wilson, 2012). Defensible space, for her, became a cluster of ideas.

On moving to the DoE in 1978 Wilson worked with Mike Burbidge on difficult-to-let estates (see Burbidge et al. 1980). Here Wilson looked at the impact of a range of initiatives: improved management practices, reduced numbers of children, introducing elderly people into tower blocks or individualised rubbish chutes. Their conclusions were that amongst the wide spectrum of age, design and layout of the estates studied, the common failing was weak housing management, inadequate maintenance and allocations: 'Real problems occur when the most vulnerable families become concentrated in the more difficult forms of housing' (Wilson 1978a: 674). Wilson felt she had uncovered a significant psychological/social problem in how social policing degraded when moving between housing typologies, particularly from terraces to large-scale estates of medium- and high-rise. For Wilson, tower blocks obviously were not working as vertical streets and needed novel criminological/social solutions. But it was a challenge that the police were unprepared for: 'gaps in crime prevention knowledge are so large that any increase in crime, to anything like US levels, would find the authorities, police included, unready' (Wilson 1981b: 38). It was this knowledge of the failings of crime statistics and existing crime-prevention techniques, as well as her reputation for reliable research, that facilitated Wilson's advocacy of Newman's concepts in both the HO and DoE.

But this was not unquestioning advocacy – Wilson's close and ongoing scrutiny of how defensible space worked in practice strengthened the appeal of the principles. Even her suspicion of over-reliance on retrofitting defensible space into troubled estates – 'modification along defensible space lines rarely resolve the social and management problems co-existing with poor design' (Wilson 1978a: 674) – led to a more practical understanding of how adjusting layouts could facilitate better management. Publishing her simplified defensible space guidelines in *Building Design* (Wilson 1981b), she attempted to clarify the confusion in the practical interpretation and application of Newman's principles. For example, elaborating on the circumstances where positive late-night uses (laundrettes, community centres and flood-lit kick about areas) could be beneficial, or where improving pedestrian short cuts could undermine any semblance of privacy or defined communal spaces. And, as ever, favourable design conditions needed to be underpinned with sustained and intensive bottom-up management. These were the approaches she endeavoured to apply in practice when she left the DoE in 1982 to work as community organiser on a difficult-to-let housing estate in Swindon.

Wilson's commitment to delivering short, accessible policy-orientated research with very practical recommendations derived from her experience of expensive but ineffectual projects that failed to provide practitioners with the kinds of advice they needed. The DoE was more interested in applicable solutions than the HO, but less able to produce robust research (interview with Sheena Wilson, 2012).

The title of British architect Barry Poyner's (1983) *Design Against Crime: Beyond Defensible Space* shows how deeply established Newman's approach became in the decade after his book was published. Despite the doubts about Newman's research methods, Poyner conceded that Newman had convinced many that the design of buildings might contribute to increasing levels of crime, and this acted as the foundation for an extensive quantity of research. Poyner's meta-review examined crime reduction research since 1972; it considered neighbourhood planning, residential burglary, wilful damage to public housing, as well as criminal activity on streets and in city centres, public transport and schools. The majority of the studies were from England, the United States and Canada, with a few from France, Australia and New Zealand, covering three kinds of literature: theoretical writings, reports of empirical research studies and guidance on security design. Each of these literatures was targeted at a specific audience of academicians or practitioners. Theoretical reviews by criminologists and designers tended to criticise both Newman's methodology and his construction of a predictive theory, leaving a negative impression and minimising the lessons relevant to practitioners. As an architect/researcher Poyner was sensitive to this, questioning whether architects, planners and others involved in decisions about the built environment could be held responsible for the rise in crime. He distinguished between social scientists (including criminologists), who were 'primarily concerned with scientific statements' and whose findings were presented as predictive descriptions, and designers whose recommendations were expressed as the prescriptive actions needed to achieve a desired affect (Poyner 1983: 4). More relevant to planning professionals were the geographical studies on mapping and distribution of crime. Poyner cited evaluations and reports of good empirical research that examined the impact of environmental factors on crime levels in a practically pragmatic way. But Poyner found the lack of conclusive findings frustrating when considering future practical applications. Similarly, the guidance and proposed interventions in the early 1980s were generic, weak and obvious ('fit stronger locks', 'improve street lighting') with little evaluation of which solutions applied to particular situations. What defensible space lacked was a collective unified interpretation that cut across the spatial geographical, planning or technical differences to generate more nuanced guidelines that the various practical built-environment disciplines could take and apply. Newman's version of defensible space had proliferated through its looseness and ambiguity, but it would take an advocate who could single-mindedly gather an evidence base to promote it, and that advocate was Alice Coleman.

Alice Coleman 'Discovers' Defensible Space

As Sheena Wilson was meeting Oscar Newman in New York City, another transfer agent emerged on the scene – Alice Coleman – a geographer aspiring to influence both policy and practice. Unlike Sheena Wilson and Anne Power (who we turn to later), she was not part of established practitioner or policy making networks. So her transfer mechanisms, while similar to Newman's (personal interactions and a polemical book), were initially unsuccessful at attracting her target governmental audience. Coleman's discovery of Newman's *Defensible Space* was serendipitous. In 1976 she was invited to spend a year as a visiting lecturer at the University of Western Ontario in Canada, undertaking field surveys of neglected urban wastelands, measuring what she termed 'dying inner city syndrome' (Coleman 1980). While on this sabbatical Coleman came across Newman's book in the university bookstore, a chance encounter she attributes to being 'a great book buyer' rather than a rational follow on, as one might expect, from her emergent thinking on conditions of urban deterioration in urban wastelands:

> Well, in the University of Western Ontario they have a very good bookshop ... And I used to go and browse down there. And I am a great book buyer. And I saw this and thought it looked very interesting. So it could be chance or it could be simply because I am that sort of person. (interview with Alice Coleman, 2008)

On reading the book, Newman's discussion of urban deterioration and his proposed solutions struck a chord with Coleman, as they were similar to the 'land use deterioration' she had noticed in British cities during her Second Land Use Survey (see Maddrell 2009; Rycroft and Cosgrove 1995). Believing that Newman's concept provided a potential solution to 'problem estates' she determined there and then to recommend it to the DoE back in England (interview with Alice Coleman, 2008).

Coleman made an initial attempt to contact the DoE when back in England in late 1976, convinced they would be interested in Newman's ideas. However, the DoE were not keen to pursue Coleman's proposal for a British study. She was disappointed that her suggestions received an unenthusiastic hearing:

> I read Oscar Newman's book when I was working in Canada, I came home with the idea that the Department of the Environment would want to know this. But they said: 'No, that's an American problem'. There wasn't a care about it here and that's why I thought I should map something to find out. That's how I started doing the work in England. Then, of course, it got me hooked and I wrote the book [*Utopia on Trial*]'. (interview with Alice Coleman, 2013[5])

The DoE were unconvinced about Newman's defensible space research for three reasons: firstly, their belief he was addressing a particularly US problem, occurring on a tiny proportion of publicly provided housing projects, home to a notoriously concentrated and disaffected (black) underclass.[6] The scale and intensity of issues being dealt with in England was completely different. Secondly, there was doubt over Newman's simplistic model of design causality. The DoE felt that long-term solutions required non-physical interventions. Coleman disagreed with the DOE's belief that:

> the cure should be socio-economic. I thought how many years have we had child allowance, how many years have we had the dole, how many years have we had this welfare? The socio-economic things were in place all the time, but they never had any effect – quite the reverse sometimes. (interview with Alice Coleman, 2013)

It seems unlikely that the DoE felt the problem (or the solution) was solely attributable to a single cause, '*all* socio-economic' as Coleman claims. There *was* already a strong suspicion of environmental determinism within the DoE as an explanatory framework. From the evidence gathered in the *Design Bulletin No 25* (1972) the DoE believed they had a robust model to explain the balance of physical and social causes of residents' satisfaction. This included design, vandalism and maintenance, as well as non-environmental aspects such as a desire to move away from an estate. Variables that did not impact on this estate satisfaction model were mostly household characteristics (employment, rent/income level, number of children) and their previous housing experiences (*Design Bulletin No 25* 1972: 27). It is clear from this research report and subsequent *Housing Building Bulletins* setting standards for the external residential environment (DoE 1976) that the DoE had great confidence in the design guidance that they were publishing to provide a suitable and thriving environment. And thirdly, the DoE were aware that researchers in the HO (notwithstanding Sheena Wilson herself) were already considering Newman's findings as part of their own research into designs that discouraged vandalism. The DoE's own report *Vandalism: A Constructive Approach* (Burbidge 1973) had proposed a mix of design and lettings approaches to reduce vandalism.

Yet Coleman was undeterred by the DoE's dismissal and in 1977 set out to gather the evidence that she felt would prove the relevance of Newman's ideas. This would consist of an extensive large-scale mapping exercise she called the *Design Disadvantage in Housing Survey*. Coleman initially selected Tower Hamlets and Southwark, the two London boroughs containing the largest number of post-war blocks of flats. Coleman had become interested in Tower Hamlets during her Second Land Use Survey, for her maps distinguished derelict land created during the Second World War from subsequent slum clearance (Coleman 1980). In Tower Hamlets Coleman found six times as much unbuilt land arising from the post-war demolition programme as from bomb damage. She felt that while rebuilding had removed unsanitary housing and reduced overall housing density, it had also increased overcrowding in the new modernist blocks (Coleman 1980, 2013a, 2013b). This is indicative of how, unlike Newman or Wilson, she approached the issue of housing from a land-use mapping and geographical survey direction. Of course this was a method Coleman was trained in due to her experience with the Second Land Utilisation Survey (see Maddrell 2009), a survey undertaken by teams of amateur surveyors (many of them geography school children) who were trained in the method of 'field mapping' (see Lorimer 2003). As Jacobs and Lees (2013) point out this was a decidedly visual science, in which field observations of land use and boundaries were inscribed onto Ordinance Survey maps. This had obvious synergies with the empirical visual assessment and locational mapping methods developed by Newman in his defensible space research.

Coleman began her ambitious research without funding in place, but quickly identified a source of support. The Joseph Rowntree Memorial Trust had funded Coleman to publish a selection of Second Land Use Survey maps. Soon after starting the *Design Disadvantage in Housing Survey*, Coleman forwarded two pages of preliminary findings to the Trust. They responded enthusiastically and on the basis of a single face-to-face meeting agreed to fund the study granting the substantial sum Coleman requested (£199,000 over five years). However, her initial Joseph Rowntree contact retired and their replacement, a sociologist, was less certain about the methodological direction of Coleman's research, preferring greater use of residents' interviews than the collation of quantitative survey data. Coleman replied to his concerns with a letter setting out the benefits of what she called a 'geographical approach' rather than applying 'sociological methods'. The research manager was interested, but wanted to have a meeting jointly with the Metropolitan Police to discuss the validity of Coleman's findings. In this meeting Coleman presented her trend lines of design disadvantage features against the proportion of blocks suffering abuse of various kinds (Coleman 1980, 2013a, 2013b). Coleman recalled:

> He was there listening and when he saw how enthusiastically the police agreed with my findings he stopped trying to make me take a sociological route and trusted me to follow a geographical one. (interview with Alice Coleman, 2013)

Eventually her survey expanded to cover 27 council estates across London, a total of over 4,099 blocks of flats, and acting as controls, another 4,172 houses. A council estate outside London was also included as a control: Blackbird Leys in Oxford. Apparently this was a case of opportunist data gathering as a member of the KCL survey team had moved into a home nearby (interview with Alice Coleman, 2013). Coleman trained her team of field researchers who went out to conduct visual surveys of her chosen field site estates:

> Well, when we were doing it, we found a lot of things in England that they [New York City] didn't have. They didn't have overhead walkways, bridges – joining the blocks. And we thought, well, we must map that. And in fact we mapped quite a lot of things, altogether about 70 different things. (interview with Alice Coleman, 2008)

The young geographer Nick Fyfe worked for Coleman at this time. Having just finished his A levels, his geography teacher, who knew Alice Coleman, introduced him to her: 'We had this little office in Surrey Street (KCL) and there were three of us and she would come in in the morning' (interview with Nick Fyfe, 2019). At the time Coleman was splitting her days between her Land Use Survey work and her defensible space work:

> she was taking her land use skills into the defensible space work and thought there was no problem … There was a lot of chat about how scientific all this was from the people involved. I remember some very bizarre conversations about how to measure the dog poo, graffiti and litter, and just how subjective it all was. (interview with Nick Fyfe, 2019)

Fyfe was later invited to Coleman's inaugural professorial lecture at KCL: 'she did this whole academic biography bit, and all her physical geography work – slides of limestone – but obviously trying to build a narrative about mapping – she didn't see a tension between limestone or land use mapping and mapping graffiti, she was completely comfortable with it' (interview with Nick Fyfe, 2019; see Figure 2.3).

Fyfe enjoyed working with Coleman, finding her extremely organised and very 'hands on', recollecting her strong presence and commitment: 'she was so sincere and committed about all this, she really bought into Oscar Newman's work'. He also remembers a colleague's amazement that she had hired a man, in that she tended to work with women. He saw her as somewhat a lone figure in the geography department and discipline, but nonetheless noted that she knew people in the House of Lords and would go off to meet Lord so and so: 'she was networking with all sorts of people'.

The nature of the evidence Coleman and her team gathered and the techniques she used to present her arguments were dictated not only by her own epistemological and disciplinary leanings, but more prosaic limitations. Unlike Newman's or Wilson's research, Coleman could make minimal use of locational crime figures. Her access to crime data improved when, following the successful meeting with Joseph Rowntree and the Metropolitan Police, Coleman approached the Chief Police Commissioner Sir Kenneth Newman, who provided her with

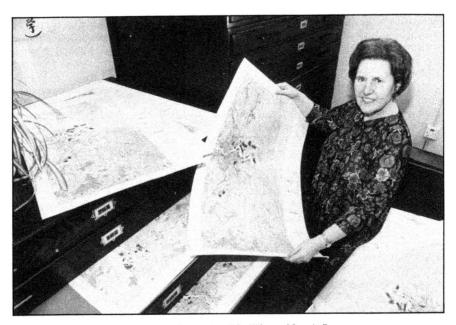

Figure 2.3   Alice Coleman and her maps. (Permission: John Wiley and Sons Ltd)

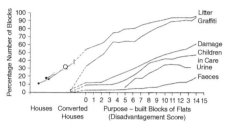

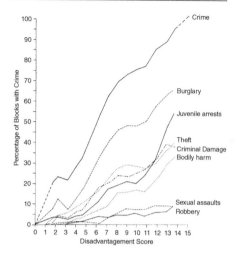

Trend lines for the six test measures in houses (left), converted houses (left-centre) and purpose-built blocks of flats (right). The houses are divided into age groups: circles for inter-war, squares for post-war, and triangles for pre-1914. in each case black symbols denote houses with, front garden fences and gates and white symbols denote houses lacking one or more of these features. The converted houses are divided into two groups. Those in Queen's Park (Q) are smaller and less littered than those in the inner city (I). Purpose-built flats are divided into classes with different disadvantagement scores from 0 to15. As their trend lines reflect the combined effect of all 15 designs they are generally smoother than those for individual designs. The graph as a whole approximates to the theoretical growth curves in Fig.8.

Trend lines for seven classes of crime in 729 blocks of flats in the Carter Street Police Division of Southwark. Crime increases as design disadvantagement worsens. No crimes were reported from the blocks with zero disadvantagement score, but the average rate for blocks with scores of 13, 14 and 15 exceeded one crime per five dwelling (1980 figures).

Figure 2.4   Coleman's trend lines for social malaise and crime. (Source: Coleman, 1985a, p. 127 [Figure 8] and p. 172 [Figure 40])

crime figures for six London police divisions. Coleman analysed this data and in February 1985 presented a report to the Metropolitan Police. The analysis compared crime data to the design assessments and measurement of 'social malaise' for the Carter Street neighbourhood in Southwark surveyed by the KCL Land Use Research Unit. This report contained initial versions of the crime trend lines (see Figure 2.4) that Coleman later used to illustrate *Utopia on Trial*.

Newman identified eight specific design variables that contributed to poor defensible space. Coleman added to these, deriving a set of measurable criteria to evaluate the design failings of individual blocks (Coleman 1985a) (see Table 2.1). Newman proposed a cluster of three variables causing anonymity arising from a large number of neighbours: the size and scale of the block and estate, the number of dwellings using the same entrance and the numbers of storeys per block. Three variables related to levels of passive surveillance: whether grounds and common areas were shared by different families; if internal corridors were enclosed and not openly visible; and the location and form of the entrance (flush entrances being preferable to set back or entrances facing away from the street). Two negative circulation factors were the presence of multiple alternative escape routes or interconnected stairs or lifts. Coleman added further factors: more than one storey per dwelling (Coleman felt flats were preferable to maisonettes which tended to accommodate families and thus a proxy for children living above ground level). Individual entrances to houses/flats were preferable to a single communal entrance.

Table 2.1   Newman's and Coleman's negative design variables.

	Coleman's design characteristics	Thresholds for harm
	Size variable	
1*	Dwellings per block	> 12
2*	Dwellings per entrance	> 6
3*	Storeys per block	> 3
4	Flats or maisonettes	Maisonettes
	Circulation variables	
5	Overhead walkways	> 0
6*	Interconnected exits	> 1
7*	Interconnecting lifts/ stairs	> 1
8*	Corridor type /Dwellings per corridor	> 4
	Entrance variables	
9	Entrance type	Communal only
10*	Entrance position	Facing into estate, distant from street
11	Doors or apertures	Open apertures
12	Piloti, garages, shops	> 0
	Features of the grounds	
13	Blocks per site	> 1
14	Access points or perimeter gates	> 1
15	Play areas	> 0
16*	Spatial organisation	Confused space

* Newman's original design variables
(Source: Adapted from Coleman 1985b: 14). See also Table 3.1 for a detailed comparison of Newman's and Coleman's design variables.

Other variables assessed the spatial organisation of the estate: the number of blocks or access points onto the site providing a distinct delineated boundary to prevent trespass. The presence of overhead walkways provided potential escape routes. Finally, a cluster of design factors regulated levels of activity: whether blocks were raised up on piloti or above garages (resulting in inactive frontages) or had play areas, which she felt acted as attractors for crime and anti-social behaviour. Coleman codified the measurement of these design features into indicators with thresholds to devise her 'design disadvantage score' (Coleman 1985a), forming the diagnostic and design basis of the remodelling process later dubbed 'Colemanisation' by the Municipal Journal (1990). Coleman's scoring system underpinned DICE (discussed in Chapter 3), but was also used by others to assess levels of physical incivilities and mental health. For example, Birtchnell et al. (1988, quoted in Halpern 1995) found strong associations between an estate design disadvantage score and levels of vandalism and residents' depression. Newman developed a similar 100-point scoring system in the Defensible Space Guidelines for Yonkers Municipal Housing Authority (Newman 1976).

From this early research on defensible space Coleman derived a hierarchy of crimes and taboos, with the weakest taboo against littering most easily broken, followed by graffiti, then more serious abuses. Like Wilson and Kelling's (1982) broken-windows theory, Coleman observed that vandalism and other abuses were more likely to occur where both littering and graffiti had a foothold. Analysing the data collected for the Design Disadvantagement Study, Coleman found this pattern repeated for each of her design variables. Supremely confident in the international read across of Newman's work, she said: 'The Land Use Research Unit at King's College London, has demonstrated that the American findings also apply to Britain' (Coleman 1985a: 560).

Utopia on Trial

Coleman had some success presenting these findings face-to-face but needed a way to communicate them to a wider audience and so began drafting a book. *Utopia on Trial* (1985a) (see Figure 2.5) was written in the form of a courtroom trial, with suspect design features and Coleman's evidence 'cross-examined'. The chapter 'Utopia Accused' dismissively condemned British post-war

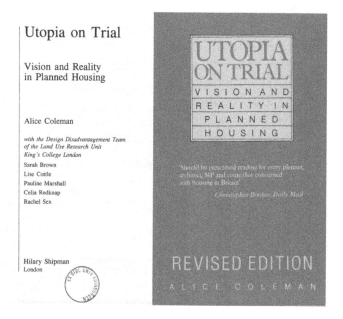

Figure 2.5   *Utopia on Trial* 1985 edition and revised 1990 edition.

housing estates as failed utopias and criticised authoritarian and paternalistic planners within the Ministry of Housing, Local Government and the DoE, while promoting 'design modification as an important new weapon in the fight against crime' (Coleman 1985a: 16). The final chapter 'Summing Up' was an admonishment of DoE researchers as 'reluctant to open their collective mind to evidence related to design' (Coleman 1985a: 181). It continued with a scathing attack on civil servants masquerading as researchers who unlike *true* researchers are not 'dedicated to truth as their highest allegiance' but 'have an interest in influencing policies and people, an activity that requires quite different mental attitudes from research' (Coleman 1985a: 181–182). *Utopia on Trial* referred infrequently to other research on the same issues, except for Coleman's admiring description of Newman's work as 'the most scientific research to date' describing it as 'a brilliant concept, quite independent in its origin and approach' (Coleman 1985a: 13).

Acknowledging the mixed reception that Newman's book received, Coleman highlighted the support for the ideas from housing officers, the police and other on-the-ground practitioners. Coleman claimed that any backlash was stirred up by those responsible for the construction of the failing blocks of flats. Regardless of poor reviews or academic criticism Coleman retained complete confidence in Newman's research that:

> proved that the relationship was not merely associative but also causal, in that if the worst values of the designs were changed to better ones, local crime rates fell, although they were rising everywhere else. (Coleman and Cross 1995: 145)

Alternatively, the support within this statement could also be interpreted as reinforcing her own position and findings.

While Coleman was writing *Utopia on Trial*, she and Newman met in New York City to discuss the basic principles behind defensible space. They then met again at KCL to talk through Coleman's early findings. They had spoken together at conferences in Canada and Holland (where they both advised on remodelling the 'notorious' Bijlmeer estate in Amsterdam). When Coleman arrived in New York City for her meeting with Newman, he said: 'frankly I'm not happy to see anyone from Britain as they rubbish my work' (interview with Alice Coleman, 2008), she reassured him that she 'valued his insight'. Coleman recounts that her book was well received by Newman, and that he indicated he liked it because, as he put it to Coleman, 'it had gotten him out from under the skirts of Jane Jacobs!' (interview with Alice Coleman, 2008). But Newman had mixed compliments for Coleman's work:

> Alice had the concept of defensible space down rather well before she arrived. What she had difficulty with were the objective physical measures. (Newman quoted in Heck 1987: 30)

Newman considered that these difficulties with calibration of measurements led to a lack of robust data on 'social malaise':

> *Utopia on Trial* does not pay sufficient attention to social factors interacting with physical causes of housing dysfunction. The thing I missed was the fit between building type and family type. (Newman quoted in Heck 1987: 30)

Newman uses this criticism to reiterate his own consideration of social factors, claiming that he carefully identified not that the built form of the high-rise was an un-liveable form, only that it was unsuitable for low-income families:

> I see high rises as quite suitable for the elderly, or for working couples, or for singles ... Alice doesn't make that distinction. (Newman quoted in Heck 1987: 31)

Having completed her design disadvantagement research and having drafted *Utopia on Trial* Coleman initially found dissemination of her findings difficult, beyond a few formal academic lectures. Looking to the popularity, reach and impact of Newman's (1972) *Defensible Space*, Coleman wished to reach a wider audience. In fact the publication of *Utopia on Trial* (1985a) as a paperback was as fortuitous as Coleman chancing across Newman's (1972) book *Defensible Space*. *Utopia on Trial* was the first publication from a new (and short-lived) publishing house, Hilary Shipman Ltd. Their intent from the start was to publish radical books, that were serious and intellectually rigorous, yet accessible to a lay audience, with the central ambition of questioning accepted views; as they termed it 'exploding sacred cows' (Macaskill 1985: para.1). There was limited mainstream newspaper interest in Coleman's research. In April 1984, Coleman was interviewed in the *Daily Telegraph* (Edmunds 1984), where she described her research and the lack of public funds to support it. She linked 18 [sic] design factors to the levels of dirt, litter and degradation and crime to be found on some estates. Hilary Macaskill (the journalist founder of Hilary Shipman Ltd) read the article and despite knowing nothing about the topic was intrigued by the forthright argument. Coleman was struggling to find an academic publisher prepared to produce a low-cost edition, so Macaskill's offer of a first paperback edition priced at less than £8 was appealing to her (interview with Hilary Macaskill, 2013).

The publishers stance challenging accepted conventions also suited Coleman and despite being from opposite ends of the political spectrum the editorial relationship was productive. Macaskill remembers an intense but collaborative editorial process prior to the publication of the first edition in May 1985. The title was a rare area of disagreement. Macaskill thought the title was catchy but insufficiently precise, believing any portrayal of urban housing estates as utopian was implausible (interview with Hilary Macaskill, 2013). However, the title proved timely, catching the mood of the moment. In an interview titled 'We were all wets when we were building Utopia' (Davie 1984)[7] Sir Hugh Casson had framed post-war housing as a utopian socialist experiment, positioning himself as anti-Thatcherite. *Utopia on Trial*'s structure as an accusative trial proved equally attention catching. Articles on the book picked up on this wordplay, for example, 'Coleman's utopia goes on trial' (*Architects' Journal* 1985) or 'It's hell for tenants in the bureaucrats' utopia' (Davie 1986), raising associations with justice, equality, objective fairness and reasoned argued debate.

Despite targeting a lay audience, it was not an easy book for a tiny publisher to promote. Initial routine pre-publication approaches resulted in subdued interest from bookshops. This disinterest rapidly reversed after a successful press splash two weeks before publication. Two articles appeared in *The Observer* and *Sunday Telegraph*. Then a long piece by Christopher Booker in the *Daily Mail* described the 'remarkable' research 'lucidly and humanly presented by Miss Coleman' (Booker 1985: para.11). Another sceptic of establishment positions, Booker was supportive of Coleman's ideas as they reinforced his own opinions stated in a BBC (1979) documentary *City of Towers*, that huge concrete housing estates were 'the greatest social disaster in Britain since the war' (Booker 1985, para.3). Booker's main thrust and final critical point however, was a warning against the dangers of central bureaucracy (particularly the DoE), who discounted what he saw as 'irrefutable conclusions', ignoring criticisms of their policies or actions. The *Daily Mail* article resulted in a flurry of media activity. At least 98 articles were listed in Hilary Shipman Ltd's press cuttings archive in the first nine months and many more subsequently (Macaskill 1986). Interviews and reviews appeared in disparate national and international journals: from *The Times Higher Education Supplement*, *The Spectator*, *New Society*, *The Glasgow Herald*, *The Australian Adelaide Review*, *Cosmopolitan*, *The Lady*, *Good Housekeeping* to *The Field*. Some of this can be explained by the media's tendency to recycle and magnify interest in a marketable commodity, but there was widespread interest and demand for a quote from the geographer Alice Coleman.

When asked what she thought about the reactions to *Utopia on Trial*, Coleman's account recalls the good with the bad:

Well, the first year it was wonderful … everybody reviewed it. I took people out, you know, famous people from the various papers and so on. And they gave me big reviews about it … And then at the end of the year I began to get some very bad reviews from people who didn't like it and just cooked up what they could say against it. For example … the architectural correspondent of *The Guardian*, wrote a most dreadful review on it … And he obviously hadn't even read it, he couldn't have read it because he couldn't have written what he had, you see. Anyway, his editor decided that it would be nice if he interviewed me. So he rang me up: 'Can I come and see you?' I said, 'no'. 'No? Why not?' [he asked]. I said 'well, because of what you have written about me, you obviously hadn't read it, you haven't got it right'. 'Oh', he said, 'you win some, you lose some, isn't it the same with you?' I said, 'No. I'm an academic and I am trying to get it as accurate as I can the whole time'. So that was that. Then, to my surprise, he wrote another review in the same paper, same author, same book, glowing. He had read it!' (interview with Alice Coleman, 2010)

There was also competition across TV and Radio. Three Radio Four programmes, *World at One*, *Today* and *Woman's Hour* wanted to interview Coleman but each demanded exclusivity. Coleman appeared on *Women's Hour*. Thames Television News filmed a first report, and the BBC planned to broadcast a debate between Alice Coleman and Anne Power (see the section 'Charges of Environmental (Architectural) Determinism') on *Newsnight* on the day of publication. During the week preceding publication, Breakfast Time TV as well as TV-AM impatiently requested interviews (Macaskill 1985). The paperback edition of *Utopia on Trial* was an immediate commercial success, with a reprint within three months. The revised edition of *Utopia on Trial*, published in 1990, book-ended the three-year lifespan of Hilary Shipman Ltd as their first and last book, and significantly it is the only publication that they continue to publish (interview with Hilary Macaskill, 2013).

In *Utopia on Trial* Coleman is highly critical of Sturman and Wilson's (1976) research, especially their finding that one socio-economic factor – child density – exerted a stronger influence than design on levels of vandalism. They had written that only if child density was under control would design 'exert a differentiating influence' (Sturman and Wilson 1976: 17). Indeed, Coleman singles out Sheena Wilson's later work and accuses her of misunderstanding Newman's premise:

> There are other examples of how people think they have mastered the gist of Oscar Newman's findings but are actually mis-interpreting them … Sheena Wilson jumped not only to the wrong conclusion that design cannot reduce crime, but also the more profound generalisation that it cannot cause it. (Coleman 1985a: 129)

Coleman is referring to Wilson's example from the regeneration of Liverpool's Angela Street estate. A ground floor area of the estate fenced off to form 'semi-public space' was soon vandalised. Wilson interpreted the speed of this damage as indicating that it was a response to the process of regeneration, not the design. This example also illustrated the ambiguity in the classification of spaces. Wilson refers to the vandalised garden as 'semi-private space' while Coleman argued that the fenced off area was 'confused space', as the garden was not visible from the windows of upper flats. So the areas Wilson considered semi-private, Coleman considered semi-public, demonstrating Ekblom's (2011) definitional inconstancies for the way that spaces were perceived and defined in these debates.

Coleman wrote that Wilson's research for the HO was 'intended to refute Newman's thesis' (Coleman 1985a: 16). But Coleman's view that she was the only promoter of Newman, defending him against systematic discrediting by a government research establishment is unfounded, as Wilson cites Newman as a principal influence, clearly stating that she aimed to explore the relevance of Newman's ideas in the UK context. In fact, Wilson's study had modest aims. Rather than establishing a unified comprehensive theory for the causes of vandalism, it set out to track the varied levels of damage across a large sample of inner London council estates and see if these variations might be explained by their layout and design. In comparison *Utopia on Trial* came to be seen as a trenchant critique of multi-unit, public housing in general and the modernist council tower block in particular (Towers 2000: 113–117).

Poyner's (1983) view of Wilson's research is far more balanced than Coleman's. Wilson's (1980) study looked at all estates with more than 100 homes in two inner London boroughs. This consisted of 285 blocks, over half of which were four or five storey gallery or balcony access blocks or nine storey slab blocks. The remainder were lower walk-up blocks, towers and a few terraces of houses. Following Newman, a similar set of characteristics were assessed: the height and size of blocks, the number of shared entrances and through routes, a classification of spaces into private, semi-private, semi-public and public. These were compared to the Local Authority housing repair records and tenancy agreements to provide a figure for child density. Newman had used the New York City Housing Authority Police Department crime records, a comprehensive and centralised source. But the records Wilson was interested in were scattered, none were computerised, requiring her to look though the paper repair chits or tenants index cards held by each housing department. Wilson undertook a visual survey, characterising the blocks and observing levels of vandalism. These relatively subjective observational measures aggregated all forms of damage (number of walls written on, smashed windows, damaged rubbish bins) as a composite block score against a four-point scale (from 1, minimal or no damage, 2, a few panes broken or some graffiti, 3, more than a few panes broken and considerable graffiti and 4, extensive boarding up, breakage and graffiti). The study analysed where damage occurred; in public or private spaces, to ground floor flats or above (lower ones were more likely to have broken windows) or to empty/occupied homes. There appeared to be little defensible space in the estates studied. The number of voids on an estate proved a good indicator for decline, with more dwellings vandalised when empty. But the most significant factor identified was child density. Wilson describes how she stumbled across the number of children per household as a critical variable, and how in retrospect had suitable records been available she would have broken this down into a more sensitive indicator of the density of boys aged between 14 and 18, as the density of boys demonstrated the clearest correlation to levels of vandalism (interview with Sheena Wilson, 2012). Yet the relationship was complicated and led to Wilson's tentative threshold of child density, with more than 3 children per 10 dwellings signifying 'high-child-density' blocks. In 'high-child-density' blocks most design typologies were found to be equally vulnerable. Where there were fewer children, rates of damage were influenced more by design factors, such as open, ungated entrances (Wilson 1980).

About the same time, another woman interested in English council estates and their problems emerged on the scene and went on to play a highly pertinent role in the mobilisation of defensible space into English housing policy and practice. Anne Power, a London-based housing activist, community organiser and housing researcher, had worked with Martin Luther King's 'End Slums' campaign in Chicago in 1966, then on community-based projects in various London boroughs, before being employed as a consultant to the DoE in 1979. Power, now considered the doyen of tenant participation, was at this time a sometime ally, but more often a foil, for Coleman's views. On moving to the London Borough of Islington in 1967 she became involved in the North East Islington Community Project, the Holloway Neighbourhood Law Centre and Holloway Housing Aid Centre, working towards the protection of tenants' rights and a new emphasis on rehabilitation not demolition. Her housing philosophy was summarised as management before construction, improvement before new building, people before property (Platt 1989: 24). Power was more conscious than Coleman of her position as a woman in a housing world where most senior positions were dominated by men, she said: 'I think it is a great loss that the women who pioneered the early philanthropic housing societies didn't manage to bring their dual emphasis on the buildings and the people to the rest of the housing profession' (Platt 1989: 25).

Anne Power recalls first meeting Coleman face to face in May 1983, on being invited to give a talk to the Royal Geographical Society on the PEP, which we discuss later, at an event jointly hosted with the Architectural Association. After she agreed to present, Power was told that they were putting Alice Coleman up against her:

> I remember she went first … All I remember is I was absolutely horrified by what she was saying, and I really went for it … It was like we had very, very divergent views. She was big time in the Royal Geographical Society, like seriously big time. (interview with Anne Power, 2018)

Of course, as Maddrell (2009) makes clear the Royal Geographical Society (RGS) was (and still is) an important way of connecting geographers with politicians, policy makers and even royalty. The RGS was/is the public face of the discipline – but a white, male, middle-class face. Inviting two female academics to the RGS, as obvious combatants, fits well with its military roots. In *The Geographical Journal* the following year, under the collective title *Trouble in Utopia*, Coleman's paper 'Design Influences in Blocks of Flats' (1984a) was accompanied by a paper from Anne Power 'Rescuing Unpopular Council Estates Though Local Management' (1984b). This was the first occasion that Alice Coleman and Anne Power were paired in print, each promoting alternative solutions to failing council-run estates, either physical modification or local management. Power recognised the cycles of regeneration and decline, arguing for constant nurturing and investment in an estate in contrast to Coleman's faith in enduring one-off design improvements:

> Anne Power was very good at going and talking to tenants and persuading them to behave better. I don't know for how long. But, I took design because it was land use, for a start, and was permanent until somebody actually changed it. (interview with Alice Coleman, 2013)

Coleman and Power were to come head-to-head again sometime later at a lunch with the head of the Nationwide Building Society:

> He obviously thought it would be a good thing to do to try to bring Alice Coleman and me together because by then it was pretty publicly known that we were very, very opposed and I think from her point of view, we were arch enemies. We had this lunch. I don't know if you've ever been to a posh banking lunch … they are ghastly … it was that kind of lunch. It was just the three of us. He unleashed the debate. Because she had me privately, she just was unbelievably offensive and awful. At the end of that, I said 'don't ever do that to me again please'. (interview with Anne Power, 2018)

Coleman and Power were pitted against each other again at another conference, Power recalls:

> There were about 500 there, they were talking about difficult council estates and defensible estates. Lo and behold I was up against Alice Coleman. They asked me if I would agree for there to be a vote at the end on whose decision was the more valid, and I said absolutely not. I'm not going to be in a competition with Alice Coleman. I'm quite happy to put my opinion across. So, I did. I went and I spoke and she spoke and it was the same performance. There was a huge question and answer session and the chair got involved and it was really, really, dynamic. That kind of discussion really plays to my strength because I don't find it at all hard to come back and come back again. Lo and behold, at the end they held a vote. So, I got about 495 votes and Alice Coleman got about 5. I can't remember the numbers actually. I'm afraid I can't remember the numbers. But it was a big gap. It was massive. She didn't like it! (interview with Anne Power, 2018)

The continual pairing and comparisons to Anne Power since the early 1980s, including Coleman's reference to Power's work in *Utopia on Trial* (1985a: 164) accentuated Coleman's alignment to the right: 'Anne Power was better looking, more dynamic, younger, more socially minded. She worked at the LSE [London School of Economics] which then implied a socialist politics – not a tough conservative' (interview with Mary McKeown, 2013; see Figure 2.6). Power became the female embodiment of left-liberal takes on council estates and Coleman the embodiment of the New Right. It is interesting that many we interviewed commented on how they looked physically, but no-one even thought to comment on what the men in the story looked like. In addition, Coleman suffered from the same treatment as Thatcher; people joked then and in interviews, 'was she really a woman', and women (mis)read as masculine, as Brown (2004) points out, often suffer punitive treatment. Coleman could

POWER TO THE PEOPLE

Management before construction, improvement before new building, people before property — this is Anne Power's housing philosophy. She talks to **STEVE PLATT** about estate management boards, tenant co-ops, and why a mass building programme would be a disaster.

Figure 2.6   Anne Power. (Source: *Roof* 1989)

be seen as one of, what Wajcman (1999) called, 'feminism's less favoured daughters' whose appearance and sexuality was closely scrutinised (see also McDowell 1997, on embodiment in the work place). In her research on gendered regimes of work, Wajcman (1999) found that organisational (in Coleman's case academic geography) structures shaped management style more than gender or personal choices. Senior women who read their organisation carefully and managed to slot in, continued to be seen as interlopers despite their entry into the upper rungs, as the institution's culture embedded and reproduced male norms and behaviours that consigned women 'out of place'. To be seen as a 'true' scientist Coleman would have needed to have demonstrated a degree of masculinity in order to be taken seriously, as did her friend Margaret Thatcher while Prime Minister.

Perhaps unsurprisingly, there was no female solidarity between Coleman and Power, despite Coleman being on *Woman's Hour*. In many ways Power fits the mould of female foil (a foil is a character who contrasts with another character, usually a protagonist) to Coleman in this story of defensible space. But we should be careful to take a 'context sensitive feminist approach' (Maddrell 2009: 9) when considering Coleman and Power, in their mutual (and mirrored) roles of protagonist and foil to each other, that attends to their differences at the same time as considering the structural similarities of their contexts and experiences. Both, despite their ideological differences, were committed to finding solutions to English housing problems, they were/are generous teachers and researchers with a lasting legacy.

Charges of Environmental (Architectural) Determinism

Coleman's epistemological position and her resultant study design was strongly positivist and her scientifically justified belief in the power of the built environment to determine social behaviour exposed Coleman to repeated accusations of environmental, architectural (physical or even social) determinism. Environmental or geographical determinism was central to the discipline of geography in the nineteenth and early twentieth centuries, explicit in the belief that the physical environment determines human behaviour. It became the foundational thought in behavioural geography, which argued that the causal mechanics for behaviour are to be found in the environment (Mitchell 2000: 17). But by the 1970s and 1980s, when Coleman began investigating defensible space, environmental determinism had not been in vogue for some time. The concept was heavily criticised for its naivety, racist and colonial past and, for many geographers, was an embarrassment to the discipline. As we show Coleman did not identify as an environmental determinist, but she was accused as such (see Meyer and Guss 2017, for a recent discussion on environmental determinism and geography).

Newman's notion of defensible space at its simplest was that crime results from the opportunities presented by the physical environment, and altering the physical environment could reduce crime. Reinforcing this, in 1971 criminologist Jeffrey (writing at about the same time as Newman) had argued that sociologists had overstated the social causes of crime and for the most part ignored environmental determinants. This, and Jeffrey's position that biological causes had also been neglected, garnered hostility from criminologists. Newman, likewise, was accused of environmental determinism for his criticism that the inhuman scale of public housing projects caused their high crime rates. To blame the design of tower blocks for having criminogenic potential, smacked of architectural determinism. The term 'architectural determinism' was first coined by a planner Maurice Broady (1966) and used (usually in a pejorative way) to describe an exaggerated belief in the extent that the design of an environment could control the way that individuals behave. Coleman was also charged with this, but as Markus complains, it was:

> deemed sufficient to damn a research project or an idea without any further analysis of the results or consideration of the methods and theories used; 'determinist', 'positivist' and 'reductionist' are common examples used to discredit work, often by critics who have but the faintest understanding of the philosophical and scientific roots of the words. (Markus 1988: 10)

Coleman's critics detected deterministic tendencies in the scientific claims that helped her get funding from Thatcher for the DICE project (see Chapter 3). Yet as Jacobs and Lees (2013) point out, Coleman's geographical background and the debate within the discipline had imbued her with a keen awareness of environmental determinism. She addressed accusations of determinism early on in *Utopia on Trial*: 'We are not dealing with determinism ... Bad design does not determine anything, but it increases the odds against which people have to struggle to preserve civilised standards' (Coleman 1985a: 83). Similarly, Coleman is clear about the contribution of design being secondary to say, the responsibility of good parenting:

> A badly designed block does not force children to become litter louts or vandals, but if the design makes it difficult for parents to supervise them and keep away from bad company, it increases their probability of behaving anti-socially. (Coleman 1984a: 351)

Noticeably, the moralistic tone of 'civilised standards' occurs frequently throughout the book and Campkin (2013: 92) argues that it is Coleman's 'presumption of bad company that is deterministic' promoting a stereotyped view of housing estate residents. Coleman repeatedly claimed that her work was merely probabilistic, brushing off negative connotations of determinism:

> I don't think it's right to say that determinism is a bad thing. My work is not determinist. It is probabilistic. All this deterministic business, it's talk about nothing. (interview with Alice Coleman, 2010)

> *Utopia on Trial* does not say that badly designed blocks make all their tenants horrible. It recognises that people are not all the same ... If the slur mongers were educated enough to read my trend line graphs properly, they would realise they cannot be deterministic. They show that as the design worsens progressively more blocks have more social break-down indicators – typical probabilistic trends. (interview with Alice Coleman, 2008)

In fact Coleman herself levelled the charge of determinism at her detractors. In *Utopia on Trial* she branded the DoE's support of Radiant City/Garden City principles as environmental determinism. She attributes any undesirable behaviour in these types of settlements as an inevitable outcome from what she saw as the poor principles and ideological town-planning dogma at the heart of Garden Cities:

> It is sad and surprising that neither the Garden City or Radiant City had any scientific background whatsoever ... Both were based upon intuitive beliefs and prejudices ... And made contagious by sincerity, enthusiasm and attractive sketches. They were carried into action by such powerful propaganda that they have become deeply embedded in our way of thinking and are difficult to dislodge. (Coleman 1985a: 6–7)

However, Coleman is equally critical of those planners who reject determinism. She accused DoE staff of 'renouncing' deterministic models as a way of absolving themselves of the consequences of their housing designs and shifting the blame on to 'problem people' (Coleman 1985a: 19). Here Coleman's simplistic reading of town planning principles (along with her cynical reading of the DoE's motivation) falls into the trap of conflating spatial versions of determinism (architectural or environmental/geographical/climatic) with social engineering. Geographer David Harvey (1997: 2) writing on urbanism identifies a failing of modernism as 'its persistent habit of privileging spatial forms over social processes'. Harvey counters that the antidote to this spatial determinism is 'to understand urbanism as a series of fluid processes in a dialectic relationship to the spatial forms which then give rise to forms which in turn create them' (Harvey 1997: 3). The task of placemaking then becomes a process of engaging and selecting a more 'socially just, politically emancipatory, ecologically sane mix of spatial-temporal production processes' (Harvey 1997: 3).

Planning and housing disciplines have displayed an instinctive distrust of being classified as determinist, without fully understanding what this criticism means, especially design disciplines that Markus (1988) suggests have a more uncertain relationship to scientific theory. Many would aspire to producing designs that aid David Harvey's socially just production processes, whilst failing to acknowledge the limitations of the tools they have to influence and improve society. Till (1998) echoes Harvey's suspicion that spatial processes are privileged over social ones, referring to Coleman:

> To promote, say, balcony access over chronic unemployment as the cause for social unrest is symptomatic of a determinist approach to architecture [that is] extraordinarily misinformed [and] extraordinarily dangerous. Misinformed because, in its focus on architecture alone, it conveniently overlooks the wider social and political structures that contribute to the production and inhabitation of the built environment; dangerous because of the political amnesia that it thereby induces. (Till 1998: 66)

We only have to look at the recent Grenfell Tower tragedy in North Kensington to see this story played out again, exposing twenty-first century political amnesia concerning council tower blocks. But the irony of course is that despite Newman and Coleman being criticised for environmental determinism, crime prevention through environmental design has become big business as we see in Chapter 6.

Physical or environmental determinism, of course, has a long history in planning and housing with the resultant politics around this. Politicians and professionals often promote as common sense that changing buildings and neighbourhoods will automatically improve people's lives. But the evidence for direct positive effects is often hard to gather, with contested outcomes complicated by other factors. Despite Coleman claiming she was not an environmental determinist, it is clear that she believed the physical environment was a determinant of behaviour, as she said on the front inner page of her 1985 book: 'It won't do ever again for academic sociologists to say that architecture does not affect people's behaviour' (Coleman 1985a). As Anson (1987: 8) concedes: 'The determinists may argue they are not determinists at all – Coleman acknowledges that "bad design does not determine anything, it just increases the odds against which people have to struggle to preserve civilised standards" – yet their writings are packed with deterministic statements about the built environment'.

In this chapter we have shown that Oscar Newman is clearly one of Stone's (2004) 'soft' policy transfer entrepreneurs facilitating a broad diffusion across disciplines and networks, whereas Sheena Wilson can be seen as part of the 'hard' policy transfer bureaucracy. Alice Coleman, however, is a more ambiguous, complex agent in that she was not fully aligned to the hard academic/policy transfer vector. While Coleman displayed essential entrepreneurial characteristics of ambitious self-belief, persistently pursuing her ideas in the face of powerful critiques and ably broadcasting them via a host of communication modes, she never fully achieved long-term success. Coleman's version of defensible space may have been heard, but her message was not fully heeded (as future chapters describe), and indeed was subject to a barrage of rebuttals. Anne Power, on the other hand, a foil to Coleman, had her views on estate regeneration heeded. She was involved in New Labour's Urban Task Force and for the most part her message was successfully diffused and has survived.

We have shown the assemblages (see McCann and Ward 2011; Peck 2011a; Peck and Theodore 2010a, 2010b) that came together as defensible space was launched in England, including those forces that dissipated it or acted as a foil to it. The 'heterogeneous assemblage of actors' each had 'their own capacities, roles and interests' but came together more often through mediators inducing 'the linkages necessary for assemblages to cohere' (Baker and McGuirk 2017: 33). Highlighting the key transfer agents associated with defensible space being mobilised in England shows how our transfer agents differed in their motivations and beliefs, how they felt about each other and interacted with each other. We trace their movements: from one side of the Atlantic to the other, from inside to outside academia, from one governmental institution to another, for example, the HO to the DoE. This shifting of course was itself part of how defensible space was mobilised (cr. Larner and Laurie 2010). We see the transfer agents also as individuals (if connected to different institutions) seeking to maximise their own personal goals and ambitions, which are in themselves political with a small and a big P. Following the policy theorist Richard Freeman (2012), who argues that accessing policy mobilities requires a focus on constitutive practices of communicative interaction, we have looked at both oral (in meetings, conferences, events) and textual (in documents) communication, which he places as central to the production and reproduction in policy-*making*. Much of what we look at is embodied practices, for as Peck and Theodore (2010b: 17) have noted: 'policy transformations … are clearly not realized declaratively or through administrative fiat; they are also embodied practices'.

We have shown the embodied social practices of meeting and talking, from formal lectures/debates and informal lunching to initiating policy-making, to be as important as the practices around numerical research and scientific 'counting' that were used to substantiate discursive authority. Geographer, Trevor Barnes' scholarship on the emergence and spread of quantitative geography and its attendant scientific claims has attended carefully to the 'peculiar social practices of individual scientists' (2004a: 280), but despite her repute in mapping as a quantitative discipline, Alice Coleman does not feature. This is a telling omission given that Barnes (2004b: 569) says that scientific ideas, despite what they might claim, are not linked to 'a polished, distant, universal rationality', rather they are 'closely tethered to the eccentricities, complex interests, materialities and messiness of lives lived at particular times and places' (Barnes 2004b: 569). In this chapter we show all of these behaviours occurring in relation to defensible space; we also show that despite her use of quantitative science Coleman was not considered as part of the British quantitative geography 'turn'. This may be because she was a woman, but equally it could be because she originated from the other side of the discipline, physical geography. More research on this idiosyncrasy within the history of geography is needed. Coleman did not emerge out of the same, very specific, conditions of production (Barnes 2004a, 2004b) that others involved in the quantitative revolution in geography did. Regardless of her being an outlier, Coleman deserves our attention and proper inclusion in the history of the discipline.

Although it is not our main focus we also reveal the differences in context between the United States and Britain during the period in which the idea of defensible space first moved across the Atlantic. It is clear that the British transfer agents were more alert to these locational differences than Newman was. Although familiar with housing conditions across Britain Wilson was shocked at the physical and social conditions she saw in the United States when meeting Newman in New York City in the early 1970s. At that time 19% of the United States' public-housing projects were concentrated in New York City – a city in such economic decline that it was near collapse, and which was undergoing large scale urban renewal projects in response (see Zipp 2010). As Jacobs and Lees (2013: 1566) point out:

> The values and ideologies of urban 'managers' (institutional gatekeepers) were paramount in the funding of urban agendas. The notion of better-designed public housing cannot be disaggregated from the often technocratic practices of the managerial city. Urban government was preoccupied with its redistributive role, i.e., the local provision of services, benefits and facilities to urban populations but burdened by a range of intractable difficulties, the most troubling of which were crime and building deterioration. From the mid 1970s (more specifically post the 1973 Oil Crisis) onwards urban governance became increasingly preoccupied with the exploration of new ways of urban development and redevelopment (see Harvey 2007: 6–7). The diagnoses and solutions offered by defensible space fitted well with the agendas of that time and place (on New York City in the 1960s and 1970s see Bellush and Netzer 1990; Berg 2007; Brecher et al. 1993).

Central to this process of 'fit to context' was adjusting solutions as required to shifting urban paradigms. As this chapter begins to reveal, and we delve into in further detail in the following chapters, defensible space adapted as it moved from a managerialist context (1970s New York City/the United States) in which Newman's work sought to serve the improvement of public-housing projects, to an entrepreneurial context (1980s London/England) characterised by Thatcherite neoliberal policies of deregulation and privatisation (Harvey 1989). We started this chapter at a time when Bevan's once dominant inclusive vision of council homes was beginning to fray, as we leave the chapter his vision is being actively undermined. With the concept of British council homes themselves under attack, interventions like defensible space also had to prove themselves to survive.

Notes

1 Bill Hillier was then at the Bartlett, UCL, but writing for the RIBA Intelligence Unit.
2 Watching the programme now, its concentration on young pre-teen children, ignoring older youths, seems nostalgic and naively dated. One clip on the Aylesbury Estate, of a group of children running up and down access ramps and jumping over railings, illustrates the lost freedom for this age group.
3 There was though British interest in vandalism in the United States at the time (see Cook 1977).
4 The Housing Development Directorate was one of the DoE research teams working on housing. Responsible for the Priority Estates Project, Anne Power was commissioned to do work for them – see Housing Development Directorate (1980) PEP (prepared by Anne Power for the SRD).
5 Coleman reported this encounter with the DoE in very similar terms when interviewed by Jane Jacobs and Loretta Lees in 2008 (see Jacobs and Lees 2013: 1567), and again by Elanor Warwick in 2013, repeating her words from *Utopia on Trial* (Coleman 1985a: 16).
6 The racialisation inherent to Newman's notion of defensible space was an aspect that did not translate across to England, and those that mobilised and operationalised the approach were clear about how the British context was/is very different. British council estates were predominantly white until the 1980s when BAME populations finally achieved access to council housing. This shift from private to municipal tenancy occurred at the same time as the so called 'urban crisis' and solving inner city problems (especially after the 1981 riots) became a focus of local and central government (see Peach et al. 1981). Although these showed the extent that the British inner city and related challenges had become racialised, the racialisation of council estates per se in the British media and beyond was just emerging as Coleman began her studies of defensible space. But as estates began to be populated by more BAME groups racialisation did take off, especially in London where some estates became majority BAME, like the Aylesbury Estate in Southwark (see the discussion in Chapter 7 on the Aylesbury Public Inquiry around crime and design; see also Lees and Hubbard 2021).
7 During the Thatcher government, the term 'wets' was applied to opponents of her more hard-line, monetarist policies, or policies reducing the regulatory power of the state.

Chapter Three
Defensible Space Goes on Trial but Attracts Those in Power

> The academic standing of a text is an uncertain guide to its social and political impact. The substantial, largely favourable reception given to Coleman's ideas is rooted in their resonance with notions and understandings which are widely prevailing. Her pronouncements on housing feed off and are amplified by the prevailing New Right ideology which they echo. (Lipman and Harris 1988: 182)

Not long after Alice Coleman's *Utopia on Trial* (1985a) was published, the book itself was put on trial. Defensible space's mobility was resisted, less by what McCann (2013) calls 'counter-mobilities', more by vociferous critique. But this opposition did not constitute a complete 'policy failing' (see Wells 2019), as defensible space was not halted or overcome. In fact, we would question whether a policy has to be stopped to have failed. Failure or success should be judged by assessing its intended impact and lasting influence, against any unintended consequences.

There was a varied reception to *Utopia on Trial* within academic, political and professional circles, and amongst a lay audience. In this chapter we look at both the positive and negative reactions to the book and its ideas, and question whether the positive responses were less to do with Coleman's ideas themselves, and more to do with the mechanisms supporting and promoting them. We also consider the impact of Coleman's own politics and personality, how she reflected the aspirations of high-level politicians, whilst irritating and

Defensible Space on the Move: Mobilisation in English Housing Policy and Practice, First Edition. Loretta Lees and Elanor Warwick.
© 2022 Royal Geographical Society (with the Institute of British Geographers). Published 2022 by John Wiley & Sons Ltd.

alienating policy-level civil servants and coming into conflict with the normative liberal/socialist politics of many housing professionals. A feisty, outspoken
individual, Coleman's politics did not sit well with many, as Guest (1990:
20) reported:

> Alice Coleman is a controversial figure. Labour councillors have refused her access
> to their estates, militant supporters have used violence to disrupt her plans. Even
> DoE civil servants have quietly prevented ministers from seeing her.

But there was partial government and policy acceptance, for example, the production of an Audit Commission report *Managing the Crisis in Council Housing* in
1986 welcomed Coleman's ideas even if it stimulated divisive arguments.

We investigate the 1987 *Rehumanizing Housing Conference* where 75 researchers, housing managers, architects and town planners plus community
activists came together in London at the Whitechapel Art Gallery. We show
that the hidden critical undercurrent at this conference was Coleman's book,
as nearly every paper referred to Coleman. Coleman did not attend the
conference, but we discuss the criticisms that came out of it from the academic
and practitioner left, and importantly what Coleman herself thought about
it. But against this background of criticism, Coleman had some unusual pro-
motors; we look in detail at the high-level support her ideas received as her
work gained purchase. Notably, she took Prince Charles on a fieldtrip outlin-
ing her ideas, after he read *Utopia on Trial* (leading to a long-lasting friendship), and we discuss her influence on his inner circle. We also discuss her
often abrasive encounters with community architects at this time, revealing
differing views on residents' agency over design changes to the places where
they live. Finally, we focus on a significant transfer moment when *Utopia on
Trial* and Coleman's ideas were accepted by Margaret Thatcher, if not the
DoE, and how this high-level support prepared the ground for Coleman
being given £50 million to test her ideas. We consider what Coleman and
Thatcher might have had in common that gave an academic geographer such
high-level access to influence and money. The chapter also shows the politics
behind defensible space and getting funding for DICE, the power plays, persistence, determination and dependence on catching the eye of the powerful
at the right time.

Utopia on Trial in the Dock

The response to *Utopia on Trial* from the press was, for the most part, positive
and politically balanced. The mainstream press of all political persuasions was
enthusiastic, although *The Observer* called Coleman 'the scourge of the DoE'

Figure 3.1   Media coverage of Alice Coleman. (Permission: Neil Libbert)

(see Figure 3.1). Even Martin Pawley, the architectural critic of *The Guardian*, not an obvious supporter of the free-market beliefs expressed, gave the book a balanced reading, gently chiding Coleman for ignoring the impact of the sale of desirable council houses and extensive cuts in public housing expenditure on the estates that she had studied (Pawley 1985). In a 1985 review titled 'Alice in Wonderland', Peter Williams (1985: 19), then assistant director of the Institute of Housing, likewise said:

> Don't dismiss this book. Some of the analysis is weak and the arguments are displaced but the central issues are relevant to today's concerns to provide a better deal for housing consumers – tenants and owners alike ... This book gives some suggestions in respect of some problems.

The wider media coverage continued, with an ITV *World in Action* programme called *Designed for Living*, broadcast in November 1985 (*Building Design* 1985a), that discussed *Utopia on Trial* and the growing popularity of community architecture. This showed Coleman and the community architect John Thompson, then Director at Hunt Thompson Associates (HTA), walking around Lea View House estate in Hackney (interview with Alice Coleman, 2013). This programme was a trial run for a later visit to Lea View House estate with the Prince of Wales.

The architectural press initially followed the national newspapers' lead promoting *Utopia on Trial*, accompanied by several articles authored by Coleman. A four-page spread in the most-read architectural weekly broadsheet, *Building Design* (Coleman 1985b), was followed by an extract of the chapter about the

Mozart Estate in the monthly *RIBA Journal* (Coleman 1985c). However, in contrast to how Coleman was feted in the mainstream press, as with Newman, the response in architectural and housing publications was less warm, with the professions annoyed by the temerity of a geographer advising them on how to rectify their designs. The mentions of *Utopia on Trial* in the architecture and planning press became more critical throughout the autumn of 1985 (e.g. Armstrong 1985; Building Design 1985b, 1985c). A report on Westminster Council's plans to implement Coleman's recommendations on the Mozart Estate antagonistically described Coleman as 'arch-critic of housing estate design and heroine of the current municipal war against architects' (*Architects' Journal*, 1985: 8). A review of *Utopia on Trial* titled 'Polemic with statistics' appeared in *Building Design* in October 1985, claiming that Coleman's research and recommendations were so lacking in credibility that they created hostility to her cause and obscured the important discussion of how to improve estates (Ash 1985). By the time that Coleman presented her research at RIBA in January 1986 the architectural profession's response was openly hostile. In an article titled 'Alice in Blunderland' Coleman was sarcastically described as an urban geographer 'radiating old-fashioned common sense' (Gorst 1986: 2), implying that she had little understanding of contemporary architectural thought. The report of the lecture in the *Architects' Journal* (1986) described the audience as 'largely sceptical' with Coleman remaining steadfast against challenges from the floor, when exceptions to her rules on walkways, large blocks or confused space were cited from 'every corner of the hall' (*Architects' Journal* 1986: 25). In response to Coleman's lecture, Byron Mikellides, an environmental psychologist, then at Oxford Polytechnic, raised the concern that social deprivation and poverty were stronger explanatory factors for patterns of behaviour on certain estates. Both articles derisively quote Coleman as irritably responding to the idea that socioeconomic factors were more relevant: '"Poverty and unemployment" she almost shouted "are *not* as strong an influence on behaviour as design"' (Gorst 1986: 2; *Architects' Journal* 1986).

One explanation for Coleman's lack of popularity amongst architects was a belief that her analysis was myopically founded on her dislike (and misunderstanding) of modernism as a specific architectural style. Her sweeping anti-modernist judgments conflated crimes such as drug taking onto design to an extent that now seems somewhat ridiculous:

> The defective designs are all fundamental features of the architectural fashion known as the Modern Movement and since its introduction into Britain the crime rate has multiplied enormously, with particular concentration in the worst designed estates. For examples, drug problems were rare in the 1980s but now are commonplace where certain design and layout features exist. (Coleman and Cross 1995: 146–147)

Coleman was promoting what architects perceived as outmoded ideas of housing design and historically discredited environmentally deterministic views. Yet confusingly her rejection not just of modernism as a style, but modernism as a formative concept, and her call to return to the social values of the past as well as the housing designs of the past, was being presented as a symbol of Thatcher's revolutionary ideas for housing. The effects of this overt political association on the popularity (or not) of Coleman's ideas with certain groups is explored below.

As much as Coleman's environmental determinism and anti-modernism antagonised the architectural profession, her conviction that design was more significant than any of the other explanatory factors was equally disparaged by academic colleagues both in her own discipline of geography but also beyond. In addition, academic critiques identified technical weaknesses: methodological and statistical failings, and her neglect of social and economic variables. Her research findings were uninformed by contemporary housing and policy context to an extent that her practical recommendations lacked plausibility. Coleman's research was vociferously rejected by many academics who viewed it as little more than pseudo-science; 'simplistic rather than simple' (Lowry 1990: 246), 'pompous' (Smith 1986a: 244) and 'under-contextualised' (Murie 1997: 32). Coleman's lack of statistical sophistication and failure to control for background variables was widely criticised (Halpern 1995; Hillier 1986a; Smith 1986a). Repeating his earlier (1973) dismantling of Newman's statistical methods, Bill Hillier unpicked Coleman's use of correlation and trend lines in his article 'City of Alice's Dreams' (1986a: 39):

> Her method of quantification of malaise is flawed, her correlations largely illusory and her attempt to test for social factors desultory.

Hillier's piece was one of a series of articles in the *Architects' Journal* professing a balanced examination of Coleman's theories.[1] Former Greater London Council (GLC) architect, then community activist and planning adviser for Planning Aid, Brian Anson's (1986) contribution derided the 'pseudo-scientific edifice' assembled by environmental determinists of whom Coleman was argued to be a chief protagonist. Both speculate on what a 'Colemanised' estate might be like, concerned that 'territorialising' blocks would result in greater segregation without addressing the root causes of the decline in inner-city housing. They wished to expose the failings in Coleman's methods as a counterbalance to the extensive publicity *Utopia on Trial* had received and the resultant risks of implementation based on false arguments. Coleman's (1986b) reply to their criticisms was detailed, addressing each statistical point and in return rebutting Hillier's own calculations. Coleman concluded by brushing off the vociferous criticisms as inaccurate 'academic games', still confident in the warm welcome her work had received from frontline housing staff. However, a fourth article in the *Architects'*

Journal from a London Borough of Hackney housing official, in turn responding to Coleman, showed that many housing professionals remained unconvinced
that the proposed remedies could resolve the complex housing and social problems they encountered (Heaven 1986).

In a later piece in the *Architect & Surveyor*, Anson (1987) took sarcastic
exception to the views of environmental determinists and warned against blaming building design and the environment for the problems of vandalism, warning
that defensible space was inherently a racialised concept:

> As the social tensions in our society increase, the determinists will be forced to seek
> out more specific human scapegoats, (in her new book, *Utopia on Trial*, Alice Cole-
> man concedes that 'vandalism may be a more aggressive way than graffiti of hitting
> out at life's frustrations') and no doubt it will be 'scientifically' proved that it is the
> frustrated poor, the frustrated ethnic groups and the frustrated blacks, who seem
> prone to the environment 'crime' which obsesses the determinists. The danger in
> environmental determinism is that it could, ultimately prove to be a racist concept.
> (Anson 1987: 8)

Coleman's geographical research methods seemed outdated and crude (Ash
1985) when her simplistic visual coding/mapping of the space around blocks
was compared to Hillier's detailed spatial analysis techniques. Coleman's lack of
familiarity with housing management regimes was shown through her failure to
consider either who might be responsible for dropping the litter, or the caretaking
processes expected to tidy up her indicators of social malaise. She counted only
the presence of litter, but not where it occurred; and failed to consider the incidence of litter per dwelling, or per number of people using a space, or by child
density. Architecture/urban design specialist Marion Roberts saw Coleman's
indicators not as a 'barometer of breakdown in society' but 'as a breakdown
in municipal housekeeping' (Roberts 1988: 123). Academic expert on council
housing, Alison Ravetz (1986: 279) argued that beneath the 'apparently clear
and "scientific" surface' Coleman was vague about essential issues: the morphol-
ogy of the blocks surveyed, how the form of houses might differ as much as flats,
construction defects, the variations in adjacent streets or the relationship of the
estate to the wider neighbourhood. This vagueness extended to practical design
matters and the mechanics of housing policy. Coleman's research condemned
design features that architects were no longer incorporating in the homes and
estates they were now building and even her complaints against the HDD at the
DoE were out of date as it had been disbanded in 1982 (Ash 1985).

An academic specialising in social policy, Paul Spicker (1987), called for
caution in deriving any policy initiatives based on Coleman's interpretation
of her data. Spicker's detailed paper pointed out the multiple statistical problems with Coleman's methods, for example her design variables not being

independent and failing to control for the interactive influence of factors. He argued that the presence of children or pensioners might explain the outcomes, as much as design. There were other crucial omissions: overlooking the impact of tenure, housing allocation policy, the sale of council properties or increasing numbers of owner-occupiers on estates. He rejected two of her recommendations (build no more flats and abandon recent design layouts) as unsupported by her evidence and argued that the third recommendation for design modifications would be a waste of resources that could more usefully be allocated to alleviating tenants' poverty:

> Coleman's dismissal of the influence of poverty is based on an unsound method and an inadequate theoretical analysis. Her recommendations for policy are in consequence a diversion from the real needs and issues. (Spicker 1987: 283)

This warning to policy makers looking for easily applied solutions and a ready-made programme to implement is repeated by Ravetz:

> Coleman offers, instead, a 'scientific proof' of causal links between design and deviant behaviours with a consequent and fool proof programme for design remedy. It is understandable that to a fast reader, particularly if he or she is a distracted official or committee member looking for solutions to problem estates, the book should seem a model of clarity. But it is in fact a classic instance of a study that is not what it seems and its head on, pragmatic approach and abundance of figures and graphs actually mask a considerable degree of confusion. (Ravetz 1988: 155)

Similarly, the young social geographer Susan Smith, having dismantled Coleman's scientific tests and the accuracy of her 'fair and unbiased' evidence concluded that there were risks in the oversimplified solutions presented:

> She has done nothing to clarify our understanding of relationships between dwelling design and the quality of life, and her recommendations are dangerous in offering politicians and planners an over-simplistic, yet superficially appealing, panacea for the complex social problems of urban communities in an ailing economy. (Smith 1986a: 246)

Influential Supporters Rally behind Defensible Space

Having full belief in the value of her research, Coleman actively marketed her book. She undertook a round of book signings promoting *Utopia on Trial*, including at The Alternative Bookshop, Covent Garden, London, shortly before its closure in 1985 (see Figure 3.2). Coleman was photographed with Chris Tame,

Figure 3.2   Alice Coleman's Libertarian Alliance book signing. (Permission: Alice Coleman)

director of the free market civil liberties think-tank, the Libertarian Alliance. A review of *Utopia on Trial* in the Alternative Bookshop's Broadsheet (Anon. 1985b) makes their appreciation of Coleman's neoliberal political position very clear. Echoing Coleman's views on modernist architecture, the broadsheet characterises the Modern Movement as 'not merely similar to Socialism, but different faces of the same collectivist catastrophe' (Anon. 1985b: 1). Comparing Coleman to Jane Jacobs, who they describe as 'no party line, classic liberal or libertarian, more a lady of the soft centre', praising Jacobs' (1961) classic *The Death and Life of Great American Cities* the review continues:

> It is hard to think of a book that has done more good and less harm, so for us Britons to say that *Alice Coleman may be our Jane Jacobs* is the highest praise there is. (Anon. 1985b: 1, emphasis added)

Coleman had in fact already sent a copy of *Utopia on Trial* to Jane Jacobs in Canada and received a letter of enthusiastic praise in return (see Figure 3.3):

> Your book is just terrific. It is the kind of guide that is needed, literally, all over the world! ... I hope your book becomes widely read and <u>used</u> as it needs to be – which is tremendously. I'll try to see that word of it gets around here.

'Here' being Toronto, where, as Jacobs continued, a problem housing project was successfully being replaced with small-scale infill houses or small apartment blocks. Jacobs concluded her supportive letter with a complaint about battling the funding authorities over what they felt was an out-moded approach. Accusations that Coleman's work was similarly old fashioned and insufficiently modern were, however, to reappear.

The publication of *Utopia on Trial* sparked international interest and Coleman re-exported her version of defensible space through appearances at international conferences, including speaking at a Canadian conference in 1986 at the invitation of Jane Jacobs. A year later she spoke at another conference sponsored by the Ontario Ministry of Housing that gathered Oscar Newman, Alice Coleman,

69 Albany Ave.
Toronto M5R 3C2
Ont. Canada
March 14, 1986

Dear Alice Coleman,

Thanks so much for sending me <u>Utopia on Trial</u>, and
for your very kind letter.

Your book is just terrific. It is the kind of guide
that is needed, literally, all over the world! A
planner from Peking, of all places, told me a couple
of years ago that they are having the same problems
with high rises and that they simply must find a way
to return to two and three storey houses, and I've seen
just such problem projects in Yugoslavia, the Netherlands
and Germany. Of course we have them here in Canada too.
How right you are that the problem-solvers managed to do
little but create problems. I was so glad to see your
reference to John Turner's book, which to my mind has
never gotten the attention or had the influence it
deserves.

I hope your book becomes as widely read and <u>used</u> as
it needs to be --which is tremendously. I'll try to see
that word of it gets around here. Toronto is just now
planning its first attempt at re-thinking a problem
project, in this case xx by trying to knit it back into
the surrounding city, from which it had been cordoned
off by the usual bad design.   In recent years the city
of Toronto has turned from projects to infill housing,
indistinguishable from the rest of the neighborhoods into
which it xxxvxvx is set and for the most part consisting
of either houses or very small (such as three or four flats)
apartment houses, and this is working well. But what a
battle it was with the provincial and federal authorities,
who provide the money! They said it was too old fashioned!

Again, thank you more than I can say.

Sincerely,

Jane Jacobs

Jane Jacobs

Figure 3.3   Letter from Jane Jacobs to Alice Coleman. (Permission: Alice Coleman)

Jane Jacobs, John Sewell (then chair of the Metropolitan Toronto Housing Authority) and architect Rem Koolhaas to speak. The conference was predicated on an anti-modernism stance with Sewell attributing the discrediting of public housing to modernism; 'Modernism gave public housing a bad name' (Freedman 1987: para.7). Koolhaas, at that point best known as a theoretical architect and author of *Delirious New York* (1978), was the only speaker to defend modernism, not for stylistic or ideological reasons, but in reaction to thoughtless hatred of a movement that was all encompassing, and a deeper more influential project than the anti-modernists would allow (Freedman 1987). Describing high-rises as 'everyone's favourite bête-noire', Newman blamed elevators that were vandalised and became jammed, recommending renovating the lower ground floors into maisonettes with back and front yards for families with children, with single and couples above, an idea Coleman was to apply in DICE schemes. He also suggested a social mix of a maximum one-third low-income amongst middle-income households. And while high-rises were 'ideal for the elderly' he advised avoiding a mix of young and old. Koolhaas presented his proposals for the regeneration of Bijlmermeer, a large 'troubled' public-housing scheme in Amsterdam, that both Newman and Coleman had visited and had advised on. A rare public appearance from Jane Jacobs reiterated her anti-modernist position.

The Ontario conference was one where Coleman's views were mainstream, but back in the United Kingdom Coleman came up against the community architecture movement (see Figure 3.4). A recurrent theme in defensible space

Figure 3.4 Coleman vs. community architects. (Permission: Cartoon Gallery. Source: *Building Design* 14 March 1988)

discussions was communities' ownership of design changes, estate management, crime prevention or the spaces on their estates. During the 1980s the EAP and PEP (discussed in Chapter 4) were the principal public funding routes for council housing renewal delivered through collaborations between Local Authority architecture/housing departments and architectural practices, including a growing number of community architects. Community architecture was becoming a formalised movement in Britain following Charles Knevitt's use of the term in the early 1970s (see Chapter 4). The premise was for community members (estate residents) to take control of decision-making and the design process, as they were most familiar with a neighbourhood's specific problems and potentially which solutions were likely to succeed. The community architect's role was one of listener and facilitator, guiding a design process that evolved from community consensus and residents' shared vision for the design of a place:

> Our role as community architects was responding to what people wanted ... and people wanted things they knew, streets and perimeter blocks. (interview with Ben Derbyshire, 2012)

As early promoters of community participation in designing-out-crime, community architects Hunt Thompson Architects (where Ben Derbyshire worked from 1976) relied on evidence gathered through community consultation, making novel use of opinion research to uncover the causes of residents' dissatisfaction with their estates, then incorporating responses into design solutions. Their remodelling of Lea View House estate, Hackney, became a template for working with residents, with architects living on the estate and estate coordinators ensuring cooperation between the architects, residents, builders and the Local Authority. Hunt Thompson Architects applied these approaches during their later project on the Mozart Estate in Westminster (see the case study following Chapter 4). Stylistically, community architecture was as anti-modernist as Coleman (interview with Ben Derbyshire, 2012), but this is an insufficient argument for why both groups came up with a comparable version of defensible space. Both proposed a more traditional street format, more legible routes, signposted by more easily identifiable, distinctive looking blocks, better street lighting and landscaping; creating safer paths to individual front doors. This suggests that the disparity was not about design as style, but more about design as process. The two processes were very different: Coleman's was a radical surgical intervention in the external physical form, versus community architects' community-led exploration of design solutions resulting in a gradual reworking of the buildings and spaces.

In 1986 Alice Coleman and John Thompson (with Bernard Hunt, founder of Hunt Thompson Architects) arranged for the Prince of Wales to visit a number of council estates in London, significantly Prince Charles already knew both people. As Jacobs and Lees (2013) discuss, according to Coleman, the Prince of Wales, who had an emerging interest in architecture and inner-city rehabilitation, read *Utopia on Trial* on his way back from a visit to Australia. He was so impressed with what he read that he contacted Coleman and asked her to tour, with him, three of the estates she had worked on (see Figure 3.5). Public historian, Patrick Wright, recounts the event, describing it to all intents and purpose as like a geography field trip:

> One fabled day in March 1986, he (Prince Charles) boarded a battered orange minibus hired from a left-wing community group in Tower Hamlets, and journeyed through the city in a company that included geographer, Alice Coleman, architects John Thompson and Richard MacCormac, and Nicholas Falk urban planner and environmentalist who organized the trip. The party visited the notorious Aylesbury Estate in Southwark and then zig-zagged up through East London into Hackney, where the Prince alighted, boarded a more reputable-looking official car, and drove around the corner for the opening of Lea View House in Hackney, a pre-war housing estate which had been refurbished by Hackney Council's Direct Labour Organization, according to a model scheme of 'community architecture'. (Wright 2009: 297)

Coleman recalls visiting the maisonette of a cousin of a royal cook 'and seeing that litter and dog dirt had been removed from the corridor I wondered whether it really would confirm the problems I had written about. But the daughter had

Figure 3.5   Prince Charles visits the Lea View House estate in Hackney in 1986 with geographer Alice Coleman, architects John Thompson and Richard MacCormac, and urban planner Nicholas Falk. (Source: Nick Wates, www.nickwates.com)

been mugged and a fire bomb had been thrown through her kitchen window, showing the seriousness of problems beyond those analysed' (interview with Alice Coleman, 2008).

Of course, field trips or study tours for facilitated learning have long been a means through which policy ideas are circulated. As Montero (2017) says, these activities present opportunities to educate local policy makers and expand local coalitions of policy agents engaged in addressing a particular policy problem, but in addition as seen here, they can also be used to draw in high level support. Coleman excelled in what Baker and Walker (2019) call 'strategic action' in the way she built and tried to connect a network of powerful actors who were interested in the concept of defensible space. Walker (2018) has called this 'structured collaboration', and these illustrate the kind of actions (widespread publicity to seek out allies, persuasive exchanges, stage managed encounters and alliance building) that bring a group of powerful actors together and to structure and choreograph their interactions.

With all this attention it is no wonder that Coleman reflected: 'I thought politically it was going places' (interview with Alice Coleman, 2008). In fact Coleman became part of the 'shifting kitchen cabinet' established by the Prince of Wales to help him form his increasingly controversial public interventions around architecture in Britain. As well as Coleman this group included Lady Rusheen Wynne-Jones, Dan Cruickshank of the New Georgian movement and Jules Lubbock, architecture critic of the left-wing *New Statesman*. Like Coleman, Prince Charles was critical of modernist architecture, its architectural dogma and design failings. These advisors, Raines (1988: 11) noted, 'would never be allowed into the bland policy briefings set up by Government', nonetheless their ideas, and those of Coleman particularly, shaped the Prince's 'urban planning philosophy'. David Harvey (2001: 178) noted with 'distress that another member of Prince Charles "kitchen cabinet" is the geographer Alice Coleman, who regularly mistakes correlation between bad design and antisocial behaviour with causation'.[2] Prince Charles' viewpoints led to running battles with architects at the time and Coleman even came to his defence: 'poor Prince Charles isn't used to this kind of backlash. I hope he's tough enough to take it" (Raines, 1988).[3] Prince Charles, unlike Coleman, had long argued that residents should have a say and design control over their living environment. In 2014 he warned that the regeneration of council estates in London were in danger of repeating the failures of the over-sized 1960s and 1970s modernist blocks,[4] promoting, via his model village Poundbury in Dorset,[5] an alternative, neotraditional, New Urbanist version of a mid-density, mixed-use market town.

There was a more mixed response to Coleman's research on defensible space from the British government. The appeal of Coleman's ideas to government looking for a panacea to problem housing is obvious, yet an article describing

Coleman as *persona non grata* at the DoE (Bar-Hillel 1986a) documented their ambivalence to her views. Unsurprisingly considering her comments about them in *Utopia on Trial*, Coleman's relationship with DoE civil servants was not conducive or constructive:

> She had been so rude about the Social Research Division and Housing Development Directorate. They were at loggerheads. Battle lines were drawn. (interview with John Harvey, 2009)

This distrust progressed up the civil service ladder. After a 'dusty response from ministers' when *Utopia on Trial* appeared (Bar-Hillel 1986a), a tentative acceptance at the higher-level was hastened by the riots in Brixton and on the Broadwater Farm estate in London during the autumn of 1985. These disturbances prompted a conference on crime prevention, organized by the Institute of Housing in March 1986, which was reported in *New Society* and *The Guardian*, as well as in the construction press (Ardill 1986; Building Design 1986; New Society 1986). Coleman ran a workshop at the conference and Sir George Young, then Junior Environment Minister, was the keynote speaker. Young reiterated that the Urban Housing Renewal Unit (UHRU) at the DoE did not accept that signs of social breakdown such as scattered rubbish, graffiti and vandalism had a simple causal relationship with design. For Coleman's recommendations to have practical benefit they needed to be accompanied by better management (Ardill 1986). John Harvey, Head of the Estate Action Team at the DoE, recalled the depth of the bitterness against Coleman:

> You couldn't mention the name Alice Coleman to any of them because she was anathema. I couldn't understand this. I came in one day and said, 'I was at a conference and I met this woman Alice Coleman'. And they said 'What!' And I said I'd told her what we were doing, and they said 'You shouldn't have said a word to her'. 'What's going on?' They said 'You've read the book?' and I said 'No what book?' '*Utopia on Trial*'. So I read it and I thought this was interesting stuff. Then they said 'Have you read the PEP research? No? Read that DoE report'. It was Mike Burbidge's commission,[6] which says there were five factors at play in rundown estates. One was the location, one was design, one was the economic profile of people, one was the remote unresponsive management, and finally, resident involvement. (interview with John Harvey, 2009)

John Harvey's surprise at the personal hostility and his openness to Coleman's ideas was not shared by his colleagues. Coleman (1986a) complained openly in the *Architects' Journal* and planning journals of the opposition that her findings had met within the department, with the *Architects' Journal* reporting there was greater interest in Coleman's work within the HO than the DoE (Heck 1987).

Other government organisations took more notice of her findings. In March 1986 the Audit Commission published a report *Managing the Crisis in Council Housing*. The report justified the use of the overworked term 'crisis' by referring to the large proportion of substandard council-owned dwellings in England. Blaming design faults such as flat roofs, deck access, non-traditional construction methods, even 'the wrong type of housing', it attributed the high numbers of 'difficult-to-let' estates (estimated by the Commission as up to 30% of some Local Authorities' stock) to these failings. The list continued with more familiar failings: shortage of rental housing, increasing home-lessness, unrealistic rent pricing and weak management control. The report criticised design standards that favoured speed and quantity before quality, particularly combined with technical problems arising from system building. But the Audit Commission's complaints extended beyond the impact on health caused by structural defects or inadequate heating systems to more social problems:

> The high-rise, deck access, 'concrete jungle' estates have seemingly bred crime, van-dalism and loneliness in many authorities. (Audit Commission 1986: 34)

The report called on housing authorities to 'be aware of the lessons to be drawn from the design mistakes of the past' (Audit Commission 1986: 34) and to refurbish run-down estates, to correct (or at least to avoid) these historic errors in the future. It directly recommended Coleman's findings, summaris-ing her refurbishment principles and reprinting her graph of Design Disad-vantagement score against types of crime from *Utopia on Trial*. In an example of the elective affinity model of research (Young et al. 2002), successfully influencing policy by reinforcing policy makers' existing beliefs, the report concluded with a statement that Coleman's recommendations corresponded closely to those reached by experienced members of the Commission. The report proposed:

> that general application of these recommendations will benefit both tenants and ratepayers. At minimum, *Utopia on Trial* should be read by all concerned with the management of local housing; and all housing authorities should submit all improvement schemes to careful evaluation in light of these well researched recom-mendations. (Audit Commission 1986: 37)

This statement proved to be the high point of governmental commendation of Coleman's version of defensible space principles.

The 1987 *Rehumanizing Housing* Conference

The following year, in February 1987, a conference titled *Rehumanizing Housing* was held in London, as a 'direct and critical response to the work of Alice Coleman and in particular her book *Utopia on Trial*' (Bulos 1987: 11). The conference aim was to fundamentally reconsider housing practice and research. But nearly every paper either referred directly to Coleman, her writing or to other judgmental articles such as Hillier's (1986a) 'City of Alice's Dreams'. So the conference's significance in academic terms should be offset against its role as a vehicle to focus criticisms on Coleman (Lowenfeld 2008). One attendee, the researcher Ian Cooper, recalls the conference as an event unlike [any] others he had participated in. Despite having done little housing research, he attended it because of the 'constellation of people who were gathered together' (interview with Ian Cooper, 2013). Yet it was not a mainstream event, independent of academic, government or private sector sponsorship. Even its location in the Whitechapel Art Gallery referred to a legacy of alternative intellectual thinking and education, emphasising the conference organisers' countervailing, independent views. The conference was framed as a new look at the contemporary crisis in Local Authority inner-city housing stock. One theme was the intransigent problems that inundated housing. Poor housing was associated with a host of issues: poor health, vandalism, homelessness, even civil unrest. These were countered by a long list of housing initiatives, including 'decentralised, locally controlled management regimes', 'resident involvement', 'community architecture', 'participatory design' or better economic opportunities (Markus 1988: 2). But the complexity, magnitude and unyieldingness of the task seemed daunting and a considerable question hanging over the debate was the possibility of finding any definitive 'solutions' at all. Cooper characterised several of the authors as 'non-resolutionists' in that they considered these problems were to be managed, not to be resolved or solved (interview with Ian Cooper, 2013). For example, Mike Jenks' paper 'Housing Problems and the Dangers of Certainty' expounded on the fallacy of 'single or simple or once-and-for-all solutions' (Jenks 1988).

One of the conference conveners, Tom Woolley, located the conference in the front line of the struggle against environmental behaviourism. Similar debates within the social sciences, particularly between sociology and architecture, had been common during the 1970s and 1980s when most architecture schools employed social scientists amongst their teaching staff, but by the mid-1980s these embedded posts had declined rapidly. Recalling the strong body of intellectual effort gathered to 'debunk' the architectural determinism movement, when interviewed, Woolley sighed, continuing 'and Alice Coleman then comes along and just revives it' (interview with Tom Woolley, 2013). Cooper used even more militaristic terms for this struggle, describing the 'rear guard action being made against determinism' and theorists still 'fighting the same battles' a decade after Newman (interview with Ian Cooper, 2013). Cooper identified the conference as a 'point of flux', a period where the world within which housing researchers were operating was altering, where the work most of them wanted to do was becoming less fashionable and less favoured by funding. Woolley puts this more strongly:

> The line-up here is essentially the last-ditch stand of people working within architecture education trying to wave the flag of responsible social sciences as it applies to architecture. And after that we never showed that character again! (interview with Tom Woolley, 2013)

So the conference was a reaction both against the revival of architectural determinism as personified by Coleman and her belief in simplistic one-off design interventions as a long-term solution. Cooper also articulated the opportunistic way that Coleman and her work was used to signify a wider threat. The conference:

> became a crystallisation moment because it was so easy to demonise Alice. You could see here a kind of coherence, of people coming together around a demonised figure. In that sense, she was a kind of representative totem. We recognised that the political landscape was changing in ways we didn't like, but that was very diffuse. But she was a concretised representation of everything that we thought was going wrong. And not only that, she was patently wrong and I think – this is a very large claim – we [the conveners of the *Rehumanizing Housing* conference] thought that if only we could point out how wrong she was, that there might be some victory, that this might reverse this underlying malaise that we could see developing – the rise of the Right. (interview with Ian Cooper, 2013)

Woolley contended that architects' criticism was (and still is) too often personality based and fails to emerge beyond a simplistic oppositional liking or loathing, with limited reflection. This conference aimed for a more academic and reasoned level of debate. To facilitate this, papers were circulated beforehand and rather than formal presentations, speakers and participants discussed them together (interview with Tom Woolley, 2013). Despite this attempt at academic, depersonalised critique, most papers were extremely disparaging of Coleman's politics, not just her research. Several of the authors, Anson (1986), Hillier (1986a, 1986b) and Ravetz (1986), had already published highly critical articles. Nevertheless, this perhaps supports their aspiration for objectivity, as any overt intention for the character assassination of an unpopular academic would have been undermined by circulating papers in advance. Discussions were recorded and while the tapes are no longer available, they were used to revise the 14 papers gathered together into an academic book titled *Rehumanizing Housing* (Teymur et al., 1988).

The published papers covered diverse topics: housing space standards, practical and formal architectural issues, community-based interventions and housing policy. Woolley described the speakers as an 'unholy alliance' of thought and experience. While all of the papers used practical examples to strengthen the connection between practice and reflection, the 15 authors' backgrounds varied. The majority claimed academic affiliation as an umbrella term;[7] yet there was only a partial attempt to move beyond the academic/epistemological community. Only four authors considered themselves practising architects, one working for a Local Authority Direct Labour Organisation. Contributors took aspects of Coleman's argument and used them to explore their own beliefs. Malpass (1988) and Jenks (1988) discussed the return to conservative Victorian values and the free market via suburban housing designs, Roberts (1988) selected the negative consequences of failures in 'municipal housekeeping'. Others were more openly critical of Coleman. Markus (1988) cites Newman's and Coleman's work as examples of ideological, fashionable, yet empirically flawed, research studies and the essential roles of critique in testing the rigour of housing research. Hillier's (1988) seminal paper 'Against Enclosure', used axial mapping of movement through the Marquess Estate in Islington, tracking the location of building entrances and burglaries in an alternative model to positive territoriality; proposing that the more segregated a home, the more vulnerable to crime it is. Ravetz's (1988) detailed critique of Coleman's statistical methods noted her lack of attention to tenure and historical influence in UK housing. The book concluded with Lipman and Harris' (1988) openly scathing attack on Coleman, 'Dystopian Aesthetics - A Refusal from "Nowhere"'. In the chapter Coleman is said to be someone who has 'recently stumbled into positivism' and is broadcasting 'this tired, discredited approach with all the excitement of a gauche convert'. *Utopia on Trial* is full of 'unbridled speculation', 'unfettered intuition' and 'streams of consciousness' dressed up as science. One could not imagine a more savage attack on academic work.

One striking impression from the conference reporting is the extent that Coleman is present through her absence. She did not attend the conference and had no opportunity to make her case, defending her views. This resulted in a less-balanced debate than in an earlier *Newsnight* dispute with Anne Power, as Cooper recalls: 'there were those who spoke from her position because there was really overt antagonism, which isn't so clear from the book' (interview with Ian Cooper, 2013).

But Woolley confirmed that while Coleman was not at the event her supporters were:

> somebody who turned up at the conference and really had a go at us on the basis that Alice Coleman was next to Jesus and who were we to dare to criticise such a wonderful person? We definitely felt like we were putting our heads above the parapet. (interview with Tom Woolley, 2013)

When the conference book was published and advertised as a 'balanced overview' (see Figure 3.6), the *Architects' Journal* asked both Anne Power and Alice Coleman to respond. Again (post Newsnight) the pairing was set up as adversarial, expecting two opposing views. Power's review supported the general sentiment of the conference, that policies and practices needed to change to achieve more humane housing. Greater influence should be given to the powerless, alienated and ignored individuals living in social housing.[8] 'Rented housing to suit the customers rather than the politicians, professionals and bureaucrats whose needs and views tend to dominate' (Power 1986: 30). Tellingly Power's housing occupiers are no longer council tenants, or even residents, but 'customers', with an unacknowledged reflection of Coleman's belief in the oppressive impacts of bureaucratic domination.

Despite a shared distrust of top-down government interventions, Power considered the *Rehumanizing Housing* conference as presenting a powerful case for rejecting the link between design and social problems. She stressed the essential role of good management, citing the multifaceted role of caretakers, not just as cleaners, but undertaking the informal repair and policing responsibilities on estates. The tone of Power's piece was measured, warning

Figure 3.6   Advert for the book *Rehumanizing Housing*. (Permission: Simon Harding. Source: *High Rise Quarterly*, May, 1988)

against 'carping antagonism, blind certainties and above all expensive design prescriptions' (Power 1986: 30). But she was unable to resist a veiled reference to Coleman's research methods. Power concluded that design modification had 'gladiator-like protagonists' who were not providing suitably holistic answers; and until this was understood researchers would 'continue to generate research grants to count KitKat and crisp wrappings' (Power 1986: 30–31).

More combatively, Coleman's response launched into a point-by-point, page-by-page rebuttal of the criticisms directed at her. She countered each of the points against her, with varying degrees of ridicule. Dismissing Lipman and Harris' paper as '42 abusive items including self-contradictory charges of positivism and negativism, or naivety and ingenuousness found compatible with ruthless manipulation' (1988: 31), she accuses her critics of lacking intellectual thoroughness. Her assailants' thinking is 'emotive', Roberts 'seems confused'. Hillier is accused not just of 'fabricating charges' about her aims, but devising 'bogus arithmetic', and Ravetz of promoting false assumptions about Coleman's data. This is overly harsh as Ravetz (1988) in fact concludes her careful statistical criticism of Coleman's methods with the measured suggestion that the truth probably lies somewhere between the extremes of data interpretation.

Coleman accused the book's editors of being blinkered by their political views, unjustly seeing her work as an attack on socialism. She criticised the book as myopic 'suffering from being too much a product of like minds and not the wider debate it claims to be' (Coleman 1988: 30). Following Power's more general review, her brusque replies read more as an angry personal spat than a balanced academic argument. This thrilled the publishers of the *Architects' Journal* and *Utopia on Trial*, who welcomed the disputatious nature of the reviews and articles as good publicity (interview with Hilary Macaskill, 2013). Transferability is determined by the ability of the transfer agent, here Coleman, to convey her idea of defensible space, independently of its reception. The transfer agent may have a positive image (e.g. Khirfan et al. 2014) or a less positive one (e.g. Bok 2014), as did Coleman, resulting in respectively the stimulation or discouragement of adoption of their opinion.

Catching Thatcher's Eye

During the 1980s perceptions of council housing were polarised by Margaret Thatcher's 'right-to-buy' scheme, instigated just months after her election and resulting in the sale of over 1.34 million council homes in the first decade after it was enacted in October 1980 (Lees and White 2020). Proving extremely popular amongst voters, the initiative rapidly siphoned off the more desirable council properties (Meek 2014). Yet even with substantial discounts, homes on unpopular estates were hard to sell, influenced by an estate's poor reputation, or high levels of crime and vandalism. Simultaneously the press covered stories of deprived, unstable populations living in neglected, poorly maintained inner-city estates and dramatized housing-related events such as the Broadwater Farm estate riots. This concentration on the extremes (the best and the worst of social housing) aided the decline in the perceptions of council-run housing, reinforcing a simplistic division of housing into 'bad' public sector and 'good' private ownership (Heaven 1986; Malpass 1988). Aneurin (Nye) Bevan's post-war ambition that council housing should be of a calibre to be the housing choice for all, had declined to a situation where the public housing sector had become the preserve of 'problem' families. Local Authorities were left with a housing stock tainted by the view that council housing was for those with no other option, or those too poor to buy out of it. Solutions were needed to dramatically transform unpopular estates, if only to maintain sufficient desirable, saleable properties for the Prime Minister's flagship 'right-to-buy' policy (see Chapter 4 where we discuss the rise and fall of council housing in more detail).

Following the mixed governmental reception of *Utopia on Trial*, Coleman proactively set out to attract the attention of those in sufficiently powerful positions to promote her research. Here we see Coleman mobilise her discursive and political connections and articulate her version of defensible space: 'the concept of articulation denotes the political-cultural work that is done to mobilise both meanings and people in order to realise a project' (Clarke et al. 2015: 55). Her publisher recalled Coleman indiscriminately contacting all the political parties, and that she was prepared to talk to anyone who she felt might provide support (interview with Hilary Macaskill, 2013). Coleman (1986b) recalled that other than a conversation with Jeff Rooker, then shadow housing minister, Labour was indifferent to her advances. But as *Utopia on Trial* became widely known, Coleman was asked to speak at several party conferences including the Green Party and the Social and Liberal Democrats (interview with Alice Coleman, 2013). In 1986 Coleman spoke to a fringe group at the Conservative Party's spring conference in Bournemouth. This session was attended by Hartley Booth, then adviser to Margaret Thatcher, who was sufficiently impressed by Coleman to buy a copy of her book.

Coleman wrote directly to Margaret Thatcher on 18 December 1987, asking for a meeting in the New Year so that she could explain why British council estates could and should be altered. On KCL letter headed paper Coleman wrote that Hartley Booth (Conservative MP) had told her that Thatcher was interested in her work on the design of council estates and asked whether she had time in her diary to meet. Her timing was very apt, as during 1987 the Prime Minister had invited housing experts to Downing Street to discuss crime prevention and housing design (Heck 1987). As a result of her letter, Coleman was invited to a brief (10 minute) meeting with the influential Conservative policy formulator Sir Keith Joseph, who was Secretary of State for Education and Science at the time. She went into that meeting armed with one of her crime graphs. Jacobs and Lees (2013) report that Joseph was clearly impressed with her, for as well as 'alerting' the cabinet to her work, he was reported in the *Geographical Magazine* as stating how much he admired the work of both Alice Coleman and planning pundit Peter Hall[9] (Anon. 1985a). Coleman was then invited to meet with Margaret Thatcher face to face in Downing Street for half an hour in January 1988. Jacobs and Lees (2013) state: 'In Coleman's narrative of her science in (policy) action she places this meeting centre stage' (p. 1573). Present were Coleman, Thatcher and two advisors (one of whom was William Waldegrave, briefly Minister for Housing and Planning between 1987 and 1988). Coleman recollects with precision and pride her own performance:

> I had been warned she might grill me, but she does not grill you. She only grills you if you do not know the answers, I had 35 man-years of research behind me and so I knew my stuff and I handled her questions. (interview with Alice Coleman, 2008)

Thatcher was well-briefed on *Utopia on Trial* in that she was asking for clarification rather than instruction. Coleman recalls this as a moment of recognition and persuasion, based on their shared commitment to a scientific model of thinking:

> Remember she was a scientist … It appealed to her because it was something which was really backed up. She had read the book, she knew what it was about and she was asking supplementary questions. (interview with Alice Coleman, 2008)

At the meeting Coleman turned to her scientific 'facts' to make her case. Unlike at her earlier meeting with Keith Joseph, Coleman did not take an actual graph with her on that day, instead she drew a graph in the air in order to demonstrate to Thatcher the scientific basis of her arguments. It was, she reflected, 'a sort of invisible visual' (interview with Alice Coleman, 2010); she remembers: 'I was at one point drawing a graph in the air and I said "I liked graphs", and she [Thatcher] joked back, so did she' (interview with Alice Coleman, 2008).

In Guest (1990) Coleman reflected: 'She (Thatcher) was very business-like, I was quite amazed by her speed of thinking'. It seems that the two women made an immediate connection, grounded in their shared belief in scientific facts as the basis for policy-making. Transferability increases when the policy (in this case concept/idea) matches the values and political objectives of the receiving actor (Minkman et al. 2018: 231). Here Coleman demonstrated and validated defensible space through 'visible, measurable, auditable performance' (Clarke et al. 2015: 56). It is helpful to keep this demonstrative performativity built on scientific values in mind when considering Coleman's interactions with other professionals, particularly the exchanges in Chapter 4 on the practising or doing of defensible space.

Coleman engaged with Thatcher in both a political and pragmatic way, applying a more empowered, Foucauldian form of agency or what Phillips (2006) calls 'agencement', in the sense of Coleman fitting or fixing defensible space to the English (and New Right ideological) context via her communicative acts. Coleman's quantitative science, her statistics, no doubt won Thatcher over:

> She (Alice Coleman) did counting, whatever she called that counting … She did it on over 65 estates, it was deeply impressive. It was huge. She was able to show the higher it [crime /litter] was, the more connections, the more common entrances, the more corridors. Whatever could possibly indicate that it wasn't a defensible space, she counted … The government was deeply impressed by this … I think because it was the sheer scale of numbers, she had numbers attached to everything which our stuff didn't. (interview with Anne Power, 2018)

When asked how she thought Coleman managed to get the influence she did, Power replied emphasising the importance of Coleman exploiting numerical geographical sensibilities:

> I don't understand how she did it… But because she does this obsessive counting, and she did this survey that's how she won them. That is how she won them … because counting was her thing and because spatial things were her thing and because she argued this was *a geographical thing*. (interview with Anne Power, 2018; emphasis added)

Coleman herself stated that the government was persuaded to give her the money for five reasons; first, the vast quantity of international research (she cited Newman's and her own) associating design with anti-social behaviour; second, that her findings associative value was supported by the predictive value; third, there was obvious practical value; fourth, 'evidence from five continents suggests that the effect of the indicated designs is a world-wide phenomenon, which must reflect something basic in human nature'; and fifth, that design improvement is a one-off method for salvaging failing housing (Coleman, in NHTPC n.d.: 15).

Coleman's geographical science may have dealt in the currency of calculation, but she also engaged in a conscious and populist discourse of persuasion (see Lipman and Harris 1988, for a critique of this). Indeed, the meeting of minds between Thatcher and Coleman could be seen as an instance of what Allen (2004: 28) refers to as 'power as seduction'. Jacobs and Lees (2013) identify the importance of Coleman sketching a graph in the air to illustrate her argument as a persuasive moment. Thatcher, according to Coleman (interview with Alice Coleman, 2008), grasped the intent immediately and convinced by Coleman's explanation, asked her what further help she needed. Coleman directly asked for money to support five years of practical trials into design improvement using her defensible space ideas on council estates. Thatcher agreed on the spot, asking when Coleman could leave her post at KCL and start as an adviser either at the HO or the DoE (Jacobs and Lees 2013).

Coleman felt that finally her research was prompting action: 'Two days later I got a letter from her, telling me to see Nicholas Ridley, the Environment Secretary. From then on it was in the bag, just a matter of waiting' (Guest 1990: 20). As Baker and Walker (2019: 17) argue, 'while data and measurement help break down and deduce an interpretation of phenomena in a particular policy space, the interaction of agents involved in policy circulation also relies on sharing conceptual understandings of policy spaces'. This describes exactly what happened when Coleman met Thatcher. Coleman was later labelled 'Thatcher's utopian thinker' and 'Thatcher's guru' (New Statesman 2012), yet outside her relationship with Thatcher Coleman's status was closer to an anti-guru. Archetypal policy gurus use charisma or persuasive expertise to promote acceptance of their ideas (Peck and Theodore 2010a), and while Coleman was undeniably recognised as an expert and leader in the field, her leadership was towards what many would consider an unfavourable direction. Like Thatcher, her leadership style was not collaborative – her strength of conviction and forthrightness was often perceived as inflexibility or even bullying – but of course many 'strong' women (including female academics) are often cast this way in a form of social misogyny. In fact, there are many stories of her kindness, and her fervour for improving the lot of others. Gurus' persuasive advocacy is open to positionality and bias. Journalist Ludovic Hunter-Tilney's interview with Coleman personified her in a very particular way: 'She walks with a cane, has a forthright bouffant hairstyle redolent of her political patron and possesses a remarkably acute memory'. Yet this Thatcherite embodiment was also apparent when Jacobs and Lees (2013) interviewed Coleman, as they walked around East London, revisiting her DICE estates, with her dressed in a fur coat and holding a Thatcher-style handbag. Margaret Thatcher's handbag had, of course, become an icon of the 1980s as a weapon yielded against her opponents, even her own ministers. 'To handbag' entered the Oxford Dictionary, meaning (of a female politician) treating a person or idea insensitively, even ruthlessly. Coleman's handbag could be seen as an example of 'always emergent' embodied practices that valorise the processes that operate before conscious thought – a 'realm of potential' (McCormack 2003: 495), as the handbag connected her in the minds of the people she encountered to Thatcher.

Subsequently both Sir Keith Joseph and then Michael Heseltine visited Coleman in the Geography Department, in the Surrey Street building at KCL, in what had been an old hotel. There she showed them her work in the departmental seminar room, where various large displays were set up showing Coleman's results and again featuring her persuasive trend lines (interview with KCL cartographer Roma Beaumont, 2011). However, the positive response from Margaret Thatcher was not the 'open sesame' that it seemed. Thatcher may have recognised that her imposition of Coleman and her ideas in a top-down way was likely to be unpopular and contrary to the DoE's established routes for deriving and testing research questions.

> I went further than the DoE in believing that the design of estates was critical to their success in reducing the amount of crime. I was a great admirer of the works of Professor Alice Coleman and I made her an adviser to the DoE, to their dismay. (Thatcher 1993: 605)

But this high-level endorsement ensured that authorisation proceeded rapidly. John Harvey (at the time lead of the DoE Estates Action team) remembers the urgent instructions to act:

> I got a phone call one day from the Secretary of State's office, 'Oh Professor Alice Coleman is here with the Minister [Nicholas Ridley], can you come up?' She was coming out from the Minister – 'Thank you for coming in to see me ... and John, you can take the Professor Coleman and explain to her how we're going to run this ... I'll fill you in later. But just go run after her'. Coleman said 'Right, it's been decided that I'm running this project, and I'll need £150 million'. So they had this plan and she [Mrs. Thatcher] said 'Alice Coleman needs the money because it's such an important social experiment, we must test it and see if it works, it's going to have a huge impact and we've got to see if it works'. So she [Coleman] said 'How much does it cost to renovate an estate, say a typical 1,000 dwelling estate?' 'Oh you know £10 million'. 'We need to have 10 of these'. So we're looking at £100 million or whatever. Figures plucked out of the air! So I was told to find £150 million for her and the idea was it would come out of the Estates Action budget. (interview with John Harvey, 2009)

The Estates Action budget was then £350 million annually (Power 1998), so it is not surprising that the £150 million Coleman requested was scaled down to a still substantial £50 million for what became known as the DICE project. At the time Housing Investment Programme (HIP) allocations allowed London boroughs to borrow money from the Estates Action fund, in a form of credit allocation, but this was not what Coleman had expected. She had hoped to have control over a capital sum, to allocate to architects and builders as she chose. Harvey had to explain the procedure to her:

> In the end, she said, 'If that's the best you can do. I think I'll complain to No 10 ... for reneging on the agreement'. I said I was trying to do things in the way it works, through the system. But eventually she agreed. (interview with John Harvey, 2009)

David Riley, who managed the DICE project (see Chapter 4) for the DoE underlined the distrust that the civil servants had for ideas being foisted upon them:

> I don't know how she got access to these people, but she did. I think it was pertinent that it [DICE] would have been killed off had it come up the usual way. The evidence wasn't strong enough to warrant that level of investment. I think it was £50 million as I remember and that might have been scaled back from her original ambitions. (interview with David Riley, 2011)

Paul Wiles, then Professor of Criminology at the University of Sheffield, who was involved in assessing the DICE initiative also identified that in addition to the increased resentment and rancour within the DoE, such highly politicised support resulted in unrealistic expectations:

> You might think it's wonderful if you catch the attention of somebody like Mrs. Thatcher. But actually what you do is create expectations that are almost impossible to fulfil. So what people were looking for from DICE was a reduction in crime that was probably not credible even if she'd been right, because she'd built this up as if this was going to solve the crime problem. (interview with Paul Wiles, 2011)

Both Wiles and Jenks complained of politicians 'jumping to simple conclusions' (Jenks 1988; interview with Paul Wiles, 2011). The political drive for reliable, immediate, large-scale solutions to contemporary housing problems, and the extent that this inevitable politicisation distorted objective scientific or academic debate was demonstrated by the route taken to obtain funding for DICE:

> I think the attraction for the Minister and Margaret Thatcher was that DICE was a universal remedy that would solve anything. Maybe with a touch of snake oil in that. (interview with David Riley, 2011)

It is a crude political caricature that Ministers do not care how or why an intervention works, as long as it is seen to make a positive difference. Yet those close to the workings of policy-making recognise the truth in Flyvbjerg's (1998: 35) observation that:

> analysis instead of acting as a foundation for intelligent policy making becomes a manipulated instrument of politics.

This acknowledges the misappropriation of evidence and analytical critiques, but also that initiatives require political alignment to gain practical traction. So having finally accessed funding through circuitous political routes, DICE was going to be judged not against Coleman's objective scientific terms but the highly critical, distrustful expectations of civil servants within the DoE.

The inevitability of politics shaping policy-making is well-known (Edwards and Evans 2011) but this chapter shows the extent that even mobilising the concept of defensible space was highly politicised at personal, organisational, as well as party political levels. In Chapter 5 we show how the evaluation of the DICE initiative took on a similar political flavour.

Thatcher, for her part, continued to praise Coleman. In March 1988 she attended the press conference to launch 'Action for Cities' at the Queen Elizabeth II Conference Centre in Westminster, in which she very publicly endorsed Coleman's work:

> I think you should also know about Estates Action which is already underway and doubtless many colleagues here will have seen the work and articles of Alice Coleman, Professor at Kings College, who we are getting to advise us extensively on estate action, housing action ... She is also very interested in designing council estates to see that you get considerable reduction in crime, because I am afraid most of these estates were designed without any thought of asking the police for their advice, of how to do the architecture to avoid areas where crime is rampant ... She is extremely successful and she has done, I think, about twenty-two estates and you are interested in her work and the Home Office is.[10]

An important task, of course, is considering the effect of Coleman's politics on how her research was received and whether it unlocked access to funding and support. We would argue that in fact it was a barrier to be overcome, as whilst it probably eased her access to influential supporters in Government (e.g. Waldegrave, Joseph and ultimately Thatcher), in general her politics set her apart from, and alienated, the communities of practice who were discussing and potentially implementing her work. Hence we see Coleman as an 'anti-guru'. Lipman and Harris' (1988) quotation at the start of this chapter highlighted the political positioning of the *Rehumanizing Housing* conference, which was an unusually intense conflict of right and left. Coleman's explicit and oppositional politicisation was shocking and confronted the normative liberal/socialist political position of many housing professionals and it is unsurprising that she did not have an easy relationship with them.

The architecture profession in particular found it difficult to articulate their confusion to this overt politicisation. The architectural researcher Ian Cooper remembers: 'it was difficult to talk about politics in schools of architecture and associated research journals of the time' (interview with Ian Cooper, 2013). Thinking about his fellow attendees at the *Rehumanizing Housing* conference, he continued, 'and what's even more unbelievable is that this was probably the most politicised group of people around at the time' (interview with Ian Cooper, 2013). Cooper and several other conference conveners and contributors, were at the time members of the New Architecture Movement,[11] to the alternative left of the architectural profession. So, while he found it refreshing to reconsider the conference as an historical alternative to the current neoliberal framing that most housing professionals now take as the status quo, reread-ing the papers, Cooper felt that the personal attacks overwhelmed any deeply considered political debate.

Many of the criticisms of Coleman went beyond an objective critique of her work. Several harsh attacks seemed to be founded on a personal dislike of her and her politics. Coleman herself could be just as outspoken and critical of her detractors, nonetheless she appears to have brushed off the personal nature of the arguments. Her publisher Hilary Macaskill was aware of the *Rehumanizing Housing* conference being set up to refute *Utopia on Trial* but believed that 'Alice was fine about it' (interview with Hilary Macaskill, 2013). Coleman was (and is) a fixedly political individual, her right of centre, free-market opinions colouring her views on most topics, from education, crime to planning. She remains deter-minedly anti-any form of government control, continuing to condemn what she sees as unnecessary planning bureaucracy (Coleman 2013a, 2013b). Yet in Ma-caskill's view Coleman's drive to gather support and publicity to widen the reach of her research was above politics and overcame her personal views (interview with Hilary Macaskill, 2013). Even while Coleman was courting the Conserva-tives, she was still prepared to talk with the Labour party about her work, despite their unresponsiveness and Coleman's own (2013b) criticism of Labour's nation-alisation policies. One can question whether Labour's indifference was due to Coleman being tainted by the Conservatives' support, and indeed even whether her ideas were at odds with their housing policy direction. Thompson (2020: 112) talks about 'a strange twist' when the Militant Tendency invited Coleman to Liverpool where she gave her 'seal of approval' for their housing policy in which traditional houses were built, later known as 'Hatton Houses' after Liverpool leader Derek Hatton. Coleman herself believed that she was able to maintain this apolitical stance. Talking to a reporter about DoE officials in 1990, she carefully pointed out that following the reprint of *Utopia on Trial* and the DICE project, DoE Civil Servants were now very helpful:

> Some did prevent me from seeing John Patten when he was Minister – but they've retired now. I am not politically orientated – you have to keep politics out of things especially when talking to tenants. We are doing this to improve their lives. (Guest 1990: 20)

Coleman's insistence that her work was above politics or at least above the political machinations of policy-making may have been merely the pose of an 'unworldly academic', as she astutely balanced politics throughout the media dissemination of *Utopia on Trial*, even referring to the campaign to obtain funding for DICE as 'the Thatcher project' (Jacobs and Lees 2013: 1572). And Coleman's opinion of DoE officials also thawed slightly following the agreement to fund DICE. In the revised 1990 edition of *Utopia on Trial* Coleman mellowed her critique of the DoE's 'design misguidance', rewording her concluding chapter to attribute blame more broadly. The DoE were no longer the 'king-pin of Britain's housing problems industry' (Coleman 1985a: 183), with the accusations now passed on to planners and the industry more generally: 'Britain's great housing/planning problems industry, have been manufacturing the problems they are supposed to be solving' (Coleman 1990: 184).

It was the certainty with which academics like Coleman promoted their position as unconditional absolutes that was (and perhaps still is) appealing to policy makers searching for reliable evidence on 'what works'. If this evidence happened to be in tune with the prevailing political direction, all the better:

> Positivism and certainties appear to characterize many conflicting claims. When these certainties coincide with political belief and dogma, as happened to a large degree in the 1980s, the brew is a potent one for public sector housing. (Jenks 1988: 53)

So, a critical constraint on the transfer of policies and ideas is their fit to the political direction of the time. Coleman acknowledged the transience of Thatcher's patronage, and how power and favour (especially the favour of higher levels of politics) can be short lived and fickle (Jacobs and Lees 2013) blaming Michael Heseltine,[12] who returned to the post of Environment Secretary in 1990, for the abrupt ceasing of support for her research:

> and she [Thatcher] was out. And old Heseltine confiscated the money she'd promised and told the police not to give me the crime figures so I was never able to write it up properly. (interview with Alice Coleman, 2013)

From Mobilisation to Practice

Importantly, the impermanence of the apparently vitriolic academic disagreements discussed in this chapter can be illustrated by looking forward a decade. In November 1998 Bill Hillier, pioneer of 'space syntax' theory at The Bartlett School of Architecture, University College London (UCL), presented a paper on cul-de-sacs and crime to an HO conference (Hillier and Shu 2000). This was reported in *Building Design* (1998) with an editorial titled 'Coleman Can't Cut the Mustard' contradicting the 'long-fashionable theories of defensible space' (*Building Design* 1998: 7). The article led to a fiery exchange of letters from readers (Bar-Hillel 1998; Randall 1998; Scott 1998), one calling for an immediate apology to Alice Coleman for misrepresenting her research as pro-cul-de-sac. This debate continued in *The Guardian* (Glancy 1998) until Hillier, again interviewed in *Building Design*, opined that Coleman had been influenced by his research unit, Space Syntax, but that the details of the dispute had moved on. Coleman, he claimed:

> said Oscar Newman was right inside the housing estates and we [SpaceSyntax] were right in the streets. It's a long time since we had any disagreements with Alice Coleman. (Hillier reported in Building Design 1998: 6)

Hillier[13] had earlier stated his 'great sympathy for what she is trying to say with her evidence', his concern being that no conclusion could be drawn from her work because of its methodological flaws (Hillier 1986a: 14), and now he openly acknowledged their convergence of opinions as the evidence for, and against, defensible space was explored further.

This chapter has demonstrated that the spectrum of reactions (from enthusiastic to antagonistic) to the concept of defensible space as it embedded and spread within England was influenced as much by the mechanisms used to promote it as resonance of the ideas themselves. A range of transfer mechanisms were used to mobilise the concept of defensible space: books, academic papers and events, where individual transfer agents introduced and positioned their version of the concept. The power of these individual encounters is illustrated through Coleman's persuasive meeting with the then Prime Minister, Margaret Thatcher, who in turn imposed their shared 'scientific' account of defensible space on the civil servants at the DoE.

Coleman's determination to continue regardless was recognised by Margaret Thatcher. What was perceived by some as inflexibility (even megalomania) appeared, in the context of these two single-minded women, as focused resolve:

> How she managed to persuade Thatcher, I don't know. I can only assume, because I've done a couple of rounds with Mrs. Thatcher on a couple of occasions, there was some kind of rivalry between them, almost a dare. 'It can't be done!' – 'I can show you how it can be done!'. (interview with Katrine Sporle, 2011)

At one level Coleman appears here as a guru for defensible space, indeed historian of municipal housing, John Boughton, calls her Oscar Newman's 'British alter-ego',[14] but in terms of the typologies of transfer agents it is fairer to characterise her as an 'anti-guru'. That is an individual who, however unlikely and despite their shortcomings, has the determination to promote and mobilise concepts. This reinforces the use of the term 'mobilise' under Ward's (2011: xxiv) reading 'in the sense that people, frequently working in institutions, mobilise objects and ideas to serve particular interests and particular material consequences'. Coleman's individual interests may have clashed with the DoE's, but her very presence as a counterfactual acted as a prompt and spur to their own research activities.

Coleman mobilised defensible space successfully and speedily by gaining the political, intellectual and ongoing support of many (opposing) stakeholders, each of whom held different perspectives and motivations. Her mobilisations were, ironically, more tactical and fluid than her epistemological position or the 'objective presentation' of her research. Academic science (social scientists included) asserts that *epistemic* knowledge is the most fixed, pure and valuable of Aristotle's three forms of knowledge (Flyvbjerg 2001). Yet this chapter has shown that within academic circles, defensible space was not a universally agreed set of principles, but influenced by disciplinary or epistemic positionality. This over-reliance on the *epistemic* view also underrates the significance of the two forms of knowledge that direct action: *techne*, knowing how to do something and *phronesis*, knowing what to do in particular circumstances (see Figure 3.7). *Phronesis*, the blend of knowledge, judgement, experience and reasoning that forms the basis for practice, relies on immersion in the decision-making context, agency and tactical understanding of constraints and opportunities. That the status of research findings is so closely embodied within the individual, and how their mobilised ideas fit within the workings/worldview of institutions is revealing.

Setting up research to impact on policy (grey arrow in Figure 3.7) is different to the research needed to influence practice or how policy encourages or constrains actions and decisions (black arrows in Figure 3.7). This diagram explains to some degree why and how defensible space had to alter from an academic concept, bending and distorting to respond to the different disciplinary criticisms, changing as it was operationalised when applied to the estates described in Chapter 4.

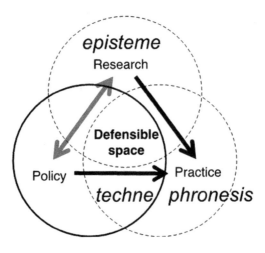

Figure 3.7   Episteme, techne, phronesis.

This chapter has concentrated on the interaction of politically motivated policy and academic desire to influence policy and action, or even resist political pressures. It is within the further process of translation of policy directives or research findings into guidance for practitioners that the theoretical clarity of research positions blurs (the dotted edged circles in Figure 3.7). At this point it is helpful to list the recommendations of the four central academics which forms the basis for much of the advice to the third group of agents – architects and planners. Table 3.1 summarises (in part) the evolution of defensible space over time with Coleman and others extending, elaborating and adding to Newman's original criteria. Additionally, it shows where the thematic interests of each academic overlap or part company. For example, Hillier is less concerned with the height of blocks than their location or 'depth' within the estate's masterplan; Power's reporting of residents' experiences is more specific about lack of privacy and opportunities for social mixing or the motivations in using spaces (and the outcomes of poorly located post boxes, bins etc.) than the routes of journeys residents may make to reach them.

Comparing Newman's, Coleman's, Power's and even Hillier's practical advice for the design of mass housing, many of their individual recommendations are less disparate than the vociferous discussion and the antagonistic opposition would lead one to expect. There *are* fundamental differences in their theorising of how space is perceived, for example, Hillier's serious, substantiated concerns over isolated estates disconnected from the street layout (Hillier 1988) leads to his and Power's opposition to Coleman's recommendation of the separation of blocks from their surroundings and enclosed spaces. Power (1998) attributed social ghettoization to this spatial segregation and in 'Against Enclosure' Hillier argued that locally enclosed urban spaces can encourage a greater vulnerability to crime, 'even though we commonly think of them as "defensible"' (Hillier 1988: 69). However, as a basic list of design criteria to be avoided, many of the recommendations of the four theorists are not mutually exclusive, nor particularly novel. The knowledge that the scale and density of large blocks can overwhelm individuals, or that secluded areas should be avoided, had long been part of designers' toolkits. Hillier's advice to avoid spaces too small to have entrances paralleled Coleman's call for individual front doors to all ground floor homes. Similarly, to Newman, Coleman's principles supported Power's suggestion that too many high-level corridors and landings dilute the numbers of people at ground level, reducing social contact. Their theoretical views started from different places but their practical advice was very similar in the end and was able to coexist in the 'real world'.

This chapter begins to show how explanations of theory are different to practical design advice. As we will see in the following chapters, policies and resultant guidance inevitably play out within a political context. For those communities of practice implementing policies in a practical attempt to improve housing, the political fit to funders or other Local Government regimes may be more

Table 3.1 Newman's and Coleman's design variables, Hillier's rules of thumb and Power's design and construction issues that make living conditions for residents difficult.

Theme	Newman (1972)	Coleman (1990: 33, 179)	Hillier (1987: 86)	Power (1998: Table 5.7 p. 96)
Block size	Number of dwellings per block (size and scale of the block).			The collective structure of the estates made individuals and families feel overwhelmed (scale).
	Number of dwellings using the same entrances.			Common entrances to blocks used by many households made it hard to keep out strangers or feel secure (design, surveillance).
		Number of storeys in a block (significant effect being the large number of neighbours rather than additional height).		Family-size units created high child densities.
		Storeys per dwelling (flats being preferable to maisonettes which Coleman perceived as a proxy for children living above ground level).		
Density of occupation				Shared services (rubbish disposal, post boxes, entrance bells, access routes) could be damaged easily and difficult to control due to their communal location (design, territoriality).
Circulation within the block	Avoid interconnected vertical routes, with each block or section of block served by only one stair or lifts.			
	Avoid interconnected exits providing multiple alternative escape routes.			

(Continued)

Table 3.1 (Cont.)

Theme	Newman (1972)	Coleman (1990: 33, 179)	Hillier (1987: 86)	Power (1998: Table 5.7 p. 96)
	Limit the number of dwellings on a corridor. Provide open visible balconies, or single loaded corridors rather than enclosed internal corridors.			Lack of sense of privacy arose from noise transmission, shared corridors and other internal spaces (design, territoriality).
Circulation in estate		Avoid all overhead walkways.		Corridors, decks, and landings on many levels led to a strong dilution of numbers of people at ground level and loss of social contact.
		Avoid blocks raised up on piloti or above garages where entrances are not visible.		
Entrance characteristics	Entrance position location and form of the entrance (flush entrances on the street preferable to those set back or entrances facing away from the street).		Orientation of facades and entrances to clarify the lines of sight and to 'mark important moments in the spatial structure'.	
		Type of entrance. Individual entrance to ground floor preferable, then communal entrances to separate parts of block rather than a single communal entrance.		
			All spawces, however small, should have building entrances open-ing on to them.	

(Continued)

Table 3.1 (Cont.)

Theme	Newman (1972)	Coleman (1990: 33, 179)	Hillier (1987: 86)	Power (1998: Table 5.7 p. 96)
Features of the grounds	Spatial organisation: reduce the degree that grounds and common areas are shared by different families. Infill or enclose leftover pieces of confused space.		Avoid clustering too many entrances in too few spaces, distribute them throughout the scheme. Avoid over enclosing spaces, except where this reflects the place of the space in the overall spatial syntax of the scheme. Avoid spaces too small to have entrances. Avoid over-hierarchisation of space, provide a range of more integrated (busier) or more segregated (quieter) zones. Avoid creating space that is empty most of the time.	Open spaces were too exposed for small groups of residents to control. Dark and secluded areas created a sense of fear and anonymity (design, surveillance).
Connection to the surroundings		Each block should have its own grounds, enclosed by a wall or a fence.	Avoid over-enclosing spaces, except where this reflects the place of the space in the overall spatial syntax of the scheme.	

(Continued)

Table 3.1 (Cont.)

Theme	Newman (1972)	Coleman (1990: 33, 179)	Hillier (1987: 86)	Power (1998: Table 5.7 p. 96)
		Number of access points onto the site should be limited to one. If more required for fire reasons, entrance/exit not located opposite each avoiding a cut through.	Link scheme visually and directly to its surroundings. Analyse existing surrounding patterns of space use, movement and encounter rates. Use this to generate schemes that relate to the wider spatial structure. Access routes should not be too deep into the scheme (more than two steps deep from the outside or an integrating core). Access routes should lead to an important destination without too many deviations.	The clear physical separation of the estates from the surrounding areas by virtue of their location, construction and tenure, evoked the notion of a ghetto (milieu).
Legibility and wayfinding			Encourage local differences in wayfinding, avoid repetition.	
Social mixing		Removal or restricted access to play areas.		Different types of households, sharing intensely communal buildings.
Design evolution based on evaluation			Preparatory studies should involve a cyclical process of design generation and systematic evaluation.	

significant than any theoretical inconsistencies of the concepts being applied. Writing up research findings consistently and persuasively may be one academic end point, but it is early in the policy journey, which is one of selecting ideas, translation, reworking and reconstitution before their application. Even at the end of the process, policy assessments rarely progress beyond superficial output evaluation. Policy evaluations tend merely to assess whether a policy output occurred, rather than providing any detailed analysis of the process, or focusing on the context, the circumstances and the mechanisms that stimulated change. Few assessments track the different routes, constraints, dead ends and the politicking required to overcome barriers for a policy to even have a chance of application, as we have here. As Kingfisher (2013: 15) says:

> Together, assemblage and translation point to the cut-and-paste processes of piecing together that are involved as policies travel up, down and sideways. It would be a mistake, however, to envision these processes as involving free-wheeling, cutting-and-pasting by sovereign agents in completely open and unconstrained environments populated by unmoored, empty signifiers ... Although fluid and unstable, translation and assemblage are also constrained. Signs and practices can be disarticulated and set off on travels in any number of directions, and policy assemblages can indeed represent cut-and-paste experimentation, but what and how things are translated, cut-and-pasted, and experimented with is not completely arbitrary.

Of course, to have impact ideas have to be tried out, and it is to the operationalisation of defensible space that we now turn.

Notes

1 The articles illustrate that *Utopia on Trial* had rapidly become ubiquitous since publication, yet by promoting these 'very personal views' the *Architects' Journal* was able to reinforce the views of Coleman's critics while appearing to be neutral. 'No conference, seminar or meeting at No 10 Downing Street on "crime and design" has been quite the same since the publication of Alice Coleman's book *Utopia on Trial*, which has been instrumental in bringing the subject to the forefront of ministerial attention. Critics of her research findings and proposals are regularly silenced by Coleman's claims on the scientific quality of her research work' (Hillier 1986a: 39).

2 Harvey (2001: 176–184).

3 *Defying tradition: Prince Charles recasts his role.* https://www.nytimes.com/1988/02/21/magazine/defying-tradition-prince-charles-recasts-his-role.html.

4 *New housing blocks risk repeating errors of the sixties, Prince Charles warns.* https://www.standard.co.uk/news/london/new-housing-blocks-risk-repeating-errors-of-the-sixties-warns-charles-9856605.html.

5 *Poundbury.* https://duchyofcornwall.org/poundbury.html.

6 Mike Burbidge was then the DoE's lead researcher at the HDD.

7   These academic roles varied from emeritus professor, chair of architecture, senior lecturers, to an early career academic working as a Local Authority research officer.

8   The rationale for community architecture was discussed during the *Rehumanizing Housing* conference, with its emergence from the rise in local tenants' management of estates establishing a more natural alignment to the theoretical position of Anne Power than Alice Coleman, whose engagement with tenants was less in-depth participatory collaboration than perfunctory consultation (see Chapter 4 on DICE).

9   Sir Peter Hall (1932–2014), planning professor at UCL.

10 *Press conference to launch 'Action for Cities'*. https://www.margaretthatcher.org/document/107188.

11 Operating between 1975 and the mid-1980s the New Architecture Movement (NAM) provided an outspoken critique of normative professional architectural structures (see http://www.spatialagency.net/database/new.architecture.movement.nam).

12 Secretary of State for the Environment May 1979–January 1983 and November 1990–April 1992.

13 Bill Hillier described his initial meeting with Alice Coleman at KCL discussing her Design Disadvantagement study data as convivial and he was pleased to be mentioned in *Utopia on Trial*. He assured us that any apparent animosity was not personal and that he 'only did a hatchet job because he'd been asked to by the DoE'. In fact, after reading 'Against Enclosure' (1987), Coleman had called Hillier up agreeing with every word. Hillier considered them later to be the 'best of friends' (personal communication with Hillier, 2006).

14 *The Pepys Estate, Deptford*. https://municipaldreams.wordpress.com/2015/08/18/a_tale_of_two_cities.

Chapter Four
Operationalising Defensible Space

In 1988 Margaret Thatcher read my report, *Utopia on Trial*, and funded me to redesign seven misery estates. A city and a housing trust commissioned two more. Threshold values and disadvantagement scores identified the defective variables and the extent of the change needed, so my method, Design Improvement Care for the Environment (DICE), was fully systematic and the anti-social activities disappeared amazingly quickly. A few small black spots were the very places where local authorities had rejected my redesign. (Alice Coleman, interviewed in *The Salisbury Review*, 2009)[1]

Alice Coleman, whose ideas have come to dominate the approach of both central government and certain local authorities to the issue of high-rise housing. (Nuttall 1988: 21)

In this chapter we look at how defensible space ideas were operationalised, that is put into practice, within England. We situate this operationalisation in the wider context of English housing policy and practice at the time but do not focus only on 'best practice' as the policy mobilities literature has tended to. Our introduction highlighted there has been little consideration of practice in the policy mobilities literature and the intention is that this chapter (and indeed the book as a whole) acts as an exemplar for future work. By practice we mean the 'doing' of making things happen, which can be explained conceptually by reference to 'assemblage':[2]

the idea of assemblage points to the ways in which policies, personnel, places, practices, technologies, images, architectures of governance and resources are brought

Defensible Space on the Move: Mobilisation in English Housing Policy and Practice, First Edition. Loretta Lees and Elanor Warwick.
© 2022 Royal Geographical Society (with the Institute of British Geographers). Published 2022 by John Wiley & Sons Ltd.

together and combined. Assemblage ... draws attention to the work of construction (and the difficulties of making ill-suited elements fit together as though they are coherent). And it makes visible the (variable) fragility of assemblages – that which has been assembled can more or less easily come apart, or be dismantled. (Newman and Clarke 2009: 9)

We begin by discussing how defensible space was operationalised through the PEP, and then the EAP, showing how both appropriated various elements from the cluster of ideas comprising defensible space. These programmes made early attempts to legitimise resident participation in regeneration, which leads us to reconsider community architecture and the Hunt Thompson Architects remodelling of the Stamford Hill estate (which Prince Charles visited). Here architect John Thompson practically applied defensible space ideas before Alice Coleman did in DICE. Thompson later distanced himself from Coleman, even if like most other architects he continues to apply defensible space principles. We then turn to DICE, then known as the Design Improvement Controlled Experiment (later amended to Coleman's altered interpretation of the acronym above) and how it played out in two London estates. These carefully selected case studies locate this chapter in the wider historical context of English housing policy during a time of flux. English housing policy in the 1970s and 1980s (see Figure 4.1) was seen as a low point for social scientists'/geographers' influence prior to the resurgence of evidence-based policy making in the 1990s. Yet the DoE (as the source of English housing policy) was a fertile location from which to conduct research that could lead to influential policy guidance and direct programmes of intervention.

DICE, known colloquially as 'Colemanisation', was the housing regeneration programme which, as Chapter 3 described, was given £50 million by Margaret Thatcher to test Alice Coleman's hypotheses of the impact of council estate design on behaviour. The money invested in this research and evaluation was an early commitment to evidence-based housing policy. As Evans (2016) says, experimentation in public policy represents a favoured and low-risk governance process that is also known to facilitate learning; to a large degree this experimentation was visible in PEP and EAP, but even more so in DICE. We explore Colemanisation from two standpoints, the academe and practice, revealing the divergent aims and ways of intervening. We discuss the setting up of the university-based DICE research team, the process for choosing the estates they focused on, and we then turn to the DICE project itself, where most of the money went to capital regeneration costs on seven DICE estates across England.

We discuss the seven DICE schemes in general and then examine two of these estates in detail (the Rogers Estate in Bethnal Green and the Ranwell Estate in Bow, Tower Hamlets), focusing on the micro-practices that played out during these major reconstruction projects. The conditions of these two estates before and after being DICE-ed are described in some detail to illustrate four enduring criticisms of defensible space as a driver for regeneration:

- physical condition is not a reliable indication of the wider cycles of regeneration and decline and hence the relative opportunities for long-term improvement;
- there is a fluctuating interrelationship between an estate and its neighbourhood;
- defensible space principles are insufficiently sensitive to an individual resident's perceptions and expectations in relation to territoriality or norms or behaviour; and
- the varied physical changes to the blocks show how difficult it is to contextualise the generic DICE defensible principles into the unique design requirements of an individual scheme.

At the end of the chapter, we look at the ways in which project teams extracted and replicated lessons from these two schemes in later projects. Throughout, we explore three themes: the impacts of policy on practice, practice on research, and research on policy. In each of these we look at the processes, mechanisms, relationships and personalities involved, to explain the resilience of the defensible space concept despite fierce criticism of it. The application of ambiguous concepts such as defensible space by urban and housing policy to justify action (and funding) is especially pertinent at the current time (and we return to this in our conclusion).

Following on from the previous chapter, we also use the opportunity to look again at our female protagonists, not only through the lens of Coleman as a central figure, but also by reflecting on her relationships with a range of women engaged in the story of defensible space and DICE. Academic rivalry between Coleman and her occasional adversary Anne Power continued, as it did between Coleman and Sheena Wilson (as a 'hidden' female researcher working within the HO/DoE). So before looking at Coleman's DICE, we return to our two other female transfer agents: Sheena Wilson and Anne Power, whose work acted as a precursor and foil to Alice Coleman's. There was also tension with the professional design and tenant management input of Coleman's research assistant on DICE – Mary McKeown, and between the various consultants employed by the DoE – Toby Taper and Katrine Sporle, and the local council employees who all forged careers off the back of what they learnt at DICE. To provide context, we trace an arc of housing strategy, as the policy focus moved from housing-led (e.g. the HIP, the PEP), through to physical design interventions (e.g. the EAP, DICE and the Estate Renewal Challenge Fund), to area/neighbourhood policies (e.g. the Single Regeneration Budget/New Deal for Communities [NDC]), and back to the recent valorisation of physical changes in estate renewal programmes. One intriguing point about defensible space is how it continues to remain relevant to housing policy despite repudiation, transformations and shifts (see Chapter Seven).

In this chapter, research idealism/policy meets real-world politics. This dichotomy of bottom-up participatory improvements against top-down design interventions can be seen as an example of Tiesdell's (2001) inevitable policy drift, as policy owners' motivations shift when theory and policy meet; what Hill (1997: 377) terms 'the inevitable messiness and incrementalism of real-world politics'. Indeed, 'DICE was not about bricks but about politics' (interview Mary McKeown, 2013). Tiesdell (2001) discusses discretion and autonomy in policy-making and implementation; actors bargaining to improve their own interests, and whether a policy is improved or distorted depending on the views of the proposed outcomes. Similarly, we discuss if difficult choices have to be made, whether it might be preferable for them to be made at the local level by implementers and *with* or *by*, rather than *for*, the intended beneficiaries of the policy. We do this by working through the operationalisation of DICE and defensible space principles on the ground in English, mainly London, council estates.

Before moving on to look at PEP and EAP in more detail it is both important and appropriate at this juncture to situate defensible space within its various architecture/design/planning systems in relation to the restructuring of the state within UK (particularly English) cities and the resultant rise and fall of council housing (especially modernist tower blocks). In the United Kingdom, the 1919 Housing and Town Planning Act and interwar push for 'homes fit for heroes' was a watershed moment in the provision of council housing for working families (if not at this time the poor). The Housing Act of 1930 encouraged mass slum clearance and councils set to work to replace slums with new build homes, trying initially to rehouse people back into the communities they were forced to vacate following the demolition of their homes. After the Second World War many slum areas remained and had been made worse by bombing during the war, in response council house building redoubled and between 1945 and 1951 1.2 million new council homes were built in the United Kingdom. By the 1960s London had an additional 500,000 new council flats added to its stock. Before the war (especially in London) council flats were characterised by the five-storey walk-up block, but this increased to 8 then 11 storeys (Towers 2000: 19). Dunleavy (1981) showed that the lions' share of British high-rise housing (five plus storeys) was built in London (36.2%). Post-war architects and planners saw slum clearance as an opportunity to enact a modern urban vision: high-rise estates within expanses of greenery, interconnected by Le Corbusier's 'streets in the sky'. In Southwark, for example, large swathes of terraced housing were bulldozed to make way for large council estates, like the Aylesbury Estate. But by the time the Aylesbury Estate was completed in 1977, this high-rise, multi-storey ideal was tarnished by failure (in part the story of defensible space told in this book) and the shift was beginning from large-scale clearances towards the preservation and rehabilitation of older housing that became associated with home improvement grants and first-wave gentrification.

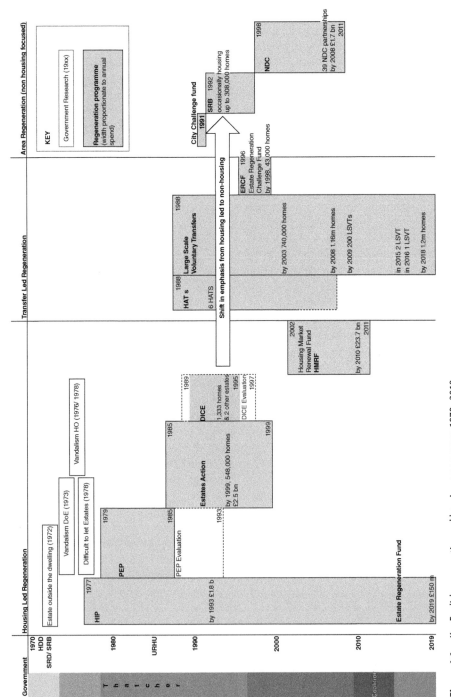

Figure 4.1 Key English regeneration and housing programmes 1979–2019.

A conceptualisation of council housing as consisting of system-built slab blocks or high-rise towers replacing serried ranks of slum terraces dates from this period, even if the reality was a more mixed picture of renewal and new construction. High-rises, while physically noticeable, and prominent in the public consciousness, continued to be a minority in public housing stock. At the same time as Alice Coleman was beginning to work on defensible space within high occupancy housing, the first national survey of high-rise Local Authority housing took place. This was the earliest systematic analysis of high-rise housing in the United Kingdom as a whole, and there has been none since, as investigations post-Grenfell have revealed. Strangely, this comprehensive census from 1983 has seldom been discussed despite its reputable researchers and important questions. The survey was led by the Housing Studies Group at South Bank Polytechnic with the help of the Institute of Housing. At the outset of their work in the early 1980s little was known about the number or distribution of high-rise tower blocks in the United Kingdom; unlike in the United States, there was no clear picture of where they were concentrated or the populations living in them. Following the DoE's definition of high-rise as six storeys or more, the survey found 4,570 high-rises in the United Kingdom (holding over 300,000 units, just over 5% of Local Authority stock), with the highest proportion in Greater London (40%) followed by the West Midlands (see Figure 4.2). Towers over 20 storeys were in the minority, with the bulk of these in major metropolitan centres, and large clusters of 10–19 storey blocks in the northern regions. Significantly, conversations with housing officers found those managing them did not consider high-rises to be any more of a problem in general than other forms of housing. The challenges housing managers did identify (insufficient investment in upkeep) would in future be exacerbated by national housing policies diluting local control over maintenance and allocations. But while the study concluded

	6–9 Storeys No. %	10–19 Storeys No. %	20+ Storeys No. %	TOTAL No. %
All Authorities	1837 (40)	2141 (47)	453 (10)	4570 (100)
West Midlands	528 (69)	216 (28)	21 (3)	765 (100)
Gtr Manchester and Merseyside	50 (13)	301 (80)	26 (7)	377 (100)
Glasgow and Strathclyde	12 (3)	187 (52)	159 (45)	358 (100)
Durham and Tyneside	31 (21)	107 (73)	8 (6)	146 (100)
Gtr London South	565 (56)	341 (34)	99 (10)	1005 (100)
Gtr London North	425 (54)	305 (28)	64 (8)	794 (100)

Figure 4.2 First national survey of all Local Authority high-rises in the United Kingdom.

that the high-rise form per se was not a problem, high-rise (in fact any block higher than three floors) was a marked problem for Alice Coleman and others. Forty years later the Ordinance Survey estimated that the number of high-rise blocks with dwellings of all tenures above six storeys exceeded 12,500 in England alone.[3] But the policy blind-spot on data on where council owned high-rises were located and who was living in them persisted.

The story of the operationalisation of defensible space in English housing policy is related to the political/policy assault on council housing from Thatcher and the New Right (see Hodkinson and Robbins 2013). This assault was both about selling properties out of council ownership into private/individual ownership, demolishing old ones and not replacing them by building new homes. Council tenants had been allowed to buy their homes since the 1936 Housing Act, and the idea of tenants having the right to own the homes they lived in was proposed in the Labour Party's unsuccessful election manifesto for 1959, but these sales had always been very limited in number. The Conservative Government's 1980 Housing Act changed this radically, by giving all council tenants the right to buy their council homes and at prices well below market value. Thatcher's 'right-to-buy' legislation saw the council housing system begin to implode with shock waves of consequences still felt 40 years later (see Hanley 2007; Malpass 2005). The number of properties managed by London councils started to shrink. In addition, the 1986 Housing and Planning Act gave councils the option of transferring all or part of their housing to another landlord (stock transfer), such as a registered social landlord, which further cemented council housing's decline (see Cole and Furbey 1994; Forrest and Murie 1988). As discussed in Chapter 3, Coleman aided the Thatcherite attack on council housing by reinforcing the ideology that it was the design of British high-rise council estates that was to blame for the social problems they experienced; regardless of the background housing shortages, rising unemployment, poor upkeep and management and underfunding of public sector housing. Coleman wrote to Thatcher in 1987, in an election year when the Conservatives pledged their commitment to expanding home ownership and incentives for property owners such as mortgage interest relief and tax exemption, while offering council tenants renovation and/or demunicipalisation. Indeed, Ortolano (2019: ch. 3) labels her 'Margaret Thatcher's housing adviser, Alice Coleman'.

More specifically, whilst extending this more general political and policy context, four points were significant in the operationalisation of defensible space. First, that investment available via the main housing programmes from the HIP through the PEP, the EAP until DICE, was constrained to individual estates (see Boughton 2018, for a summary of the funding struggles of various regeneration programmes). Second, that there was a shift in emphasis from housing-led to non-housing-led regeneration policy in the early 1990s, just as DICE was beginning. The later phases of stock transfer-led regeneration, Housing Action Trusts (HATs) and Large-Scale Voluntary Transfers (LSVTs) centred on the extensive transfer of multiple housing estates into local management either under Arms Length Management Organisations that provide housing services on behalf of a Local Authority, or housing associations. Area-based policies such as the later City Challenge Fund, Single Regeneration Budget (SRB) or NDC, pooled budgets across policy areas and funded social and physical solutions for whole neighbourhoods. Importantly, these programmes, plus Housing Market Renewal, 'took place within a uniquely favourable phase of economic and urban development' (Leather and Nevin 2013: 871). Third, there was extreme variation in the relative scales of investment, numbers of homes affected by each of the programmes and the differing periods of evaluation of these projects. The DICE evaluations although lengthy, were far shorter than the evaluation of the PEP. Finally, within the DoE there were separate directorates, each with differing remits that undertook housing research with the intention of providing practical design guidance. These operational constraints, imposed by economic circumstances and civil service protocols, set the rules within which housing regeneration functioned.

Priority Estates and Estates Action: Government's Top-down Solutions

Newman's research in New York City was funded by a broad, yet coordinated, cluster of public institutions: the New York City Housing Authority, HUD, as well as the National Institute of Law Enforcement and the US Department of Justice. Under the US Interagency Urban Initiatives Anti-Crimes programme, US government departments were working together on the issue of housing and crime. The situation in the United Kingdom was much less joined-up, with responsibility for physical crime prevention confusingly spread across several departments. So, while the HO had started to popularise a more situational crime prevention approach across national government, practical activities to reduce crime were still being addressed in a fragmented way. For example, the Department of Transport was responsible for physical design interventions such as the closure of footpaths at the rear of houses through the mechanism of Gating Orders under section 129A of the Highways Act 1980. One HO situational response focused on practical local interventions was the creation in 1978 of NACRO's Crime Prevention Unit, later expanded in 1981 by the formation of the Safe Neighbourhoods Unit (SNU). Initially funded by the GLC, the SNU operated within London boroughs until becoming an independent, non-profit organisation with a national remit in 1990 (Shaftoe 2004). During the early 1980s various Local Authorities commissioned the SNU to undertake local estate-level crime audits, promoting tenant management and community safety through estate design, use of wardens and CCTV (SNU 2009). The SNU's logo was 'policy into practice' (see Figure 4.3). They developed particular expertise in multi-storey block revitalisation and acted as consultants for the DoE's Estates Action team.

Figure 4.3   Safe Neighbourhoods Unit logo.

It is not clear if Newman ever directly approached either the DoE or the HO with his research, although, as noted, he shared platforms with their researchers at conferences. And while the research from these two departments complemented each other, it is telling that it was not until much later, following a SNU (1993) review of crime on council estates or the HO evaluation of the impact of the PEP on crime (Foster and Hope 1993), that collaborative work in this area occurred. But at this point in time in the 1980s, what is clear is that Sheena Wilson (with the external stimulus of Newman) can be seen as a key transfer agent transferring knowledge between the two departments.

Wilson moved from the HO to the DoE in 1979 looking for the opportunity to explore more rounded, practical responses to the housing problems she had encountered. Through her work with Mike Burbidge described later in this chapter, she was highly influential in shaping DoE thinking about vandalism, crime, interventions on housing estates, and in setting the foundations for the PEP. She characterised the DoE at the time as being more interested in applicable solutions than the HO, but less able to produce robust research:

> The focus moved from crime to public housing because there were so many difficulties. The whole inner urban research and the criminological research [at that time] was all about causes and blame. Then it moved to tackling, not solving; solving's top-down but tackling's bottom-up. Defensible space was about imposing solutions, whereas the process you're describing [community architecture/resident engagement] is about trying out a whole lot of different things but ultimately trying to get a bit of responsibility over to members of the community rather than coming in from outside with solutions. There is no such thing as a solution. (interview Sheena Wilson, 2012)

Wilson (directly echoing Coleman's denigration of the DoE in *Utopia on Trial*) was quite critical of the ability of government researchers to deliver 'useful research':

> I can tell you, moving from being a criminologist with the Home Office to being a social researcher for the DoE, the DoE is full of people like Alice Coleman, who use hundreds of thousands of pounds of public money to do academic research, where they bit off more than they could chew, they didn't know how to analyse it properly

and they certainly couldn't move from the analysis to policy. Government research offices were littered with people like that. (interview Sheena Wilson, 2012)

Wilson's interpretation of Newman's principles published in the *RIBA Journal* (Wilson 1981a) demonstrated her ability to translate research findings into direct advice for architects. For example, taking his principle that private space generates territoriality through two very different mechanisms – familiarity with neighbours and the ability to personalise the space outside your home – she advised carefully distinguishing real barriers from symbolic ones (e.g. a row of planter tubs or a recessed porch), which could provide 'home ground' for residents to initiate contact with their neighbours. To counter her perception that Newman's ideas were being applied too literally, she devised a new set of rules for blocks of flats where public surveillance is less effective. These covered allocation policies to restrict numbers of children and setting exceptionally high maintenance standards, with caretakers and controlled access (e.g. lockable doors) rather than merely limited access (e.g. design of entrances or layouts to serve a limited number of dwellings).

Wilson dated the increasing desire for policy to rapidly distil practical measures from academic research as starting in the early 1980s, a clear forerunner of evidence-based policy-making, but she also identified the inability of British government departments to understand the constraints of implementing their policy on the ground:

> You had these incredibly literate civil servants being asked to get their hands dirty in something they didn't understand at all. They were amenable to the idea, but they hadn't any idea what to do. I mean they weren't community activists at all, they were mandarins. (interview Sheena Wilson, 2012)

She distinguished between academics theorising about residents, and community 'agitators', noting the lack at the time of 'professional intermediaries' who could take theoretical ideas and shape them into ways of working. She welcomed a new class of professionals specialising in community engagement as stimulating essential innovation in knowledge transfer. Wilson's views here are based on her own experiences in attempting to establish a Tenants Association on a difficult-to-let estate in Swindon, aiming to apply her research findings in practice. Comparing herself unfavourably to other more grass roots organisers,[4] she attributes this failure to a mixture of personal and political naivety:

> The Tenants Association was riddled with factions and I probably wasn't charismatic enough … My background was academia so I was too analytical and not hands on enough … I wasn't a social activist. (interview Sheena Wilson, 2012)

Such self-reflection is increasingly important today in the face of an increased interest in scholar-activism in geography, especially around social housing (e.g.

Lees and Ferreri 2016; Watt and Minton 2016). In fact, Wilson felt her most useful contribution was enabling busy housing managers to explore ideas and alternative solutions. Wilson left the DoE in 1984 and was working on the Swindon estate improvement scheme when *Utopia on Trial* was published. Despite Wilson being one of the leading governmental researchers associated with estate regeneration and defensible space, having published academic papers, press articles and HO guidance during the period that Coleman was surveying her London estates, she and Coleman never met or discussed their respective positions. Because Coleman and Wilson did not meet, Coleman's condemnation of Wilson in *Utopia on Trial* could be read as Coleman's loud voice drawing out alternative views and drowning out quieter voices. Coleman's approach – far more public and utilising opinionated lectures and cajoling pronouncements – contrasted with Wilson's more modest, careful restatements and simplified advice (for example, see her articles for the *RIBA Journal*).

Wilson's research on vandalism for the HO had been prompted by a DoE paper for the HDD: 'Vandalism; A Constructive Approach' (Burbidge 1973). Wilson had moved to the DoE to collaborate with Mike Burbidge (then the HDD's lead researcher) on a far larger study *An Investigation of Difficult to Let Housing: Case Studies of Pre-War Estates* (Burbidge et al. 1980). The major outcome of the 'difficult-to-let' study was the DoE's PEP which became an influential and far-reaching housing programme. PEP ran from 1979 to 1987 and aimed to improve problem estates, using participation/representation from local communities. Based on the findings of this investigation, the DoE wanted to decentralise housing management and increase tenant involvement, believing that design and management were key factors in tenants accepting their housing. Starting in 1979, the PEP's first action was to assess 30 large estates (not only estates of high-rise flats, but also Radburn layouts and traditional pre-war cottage layouts) for the causes of their lack of popularity. In conclusion, the PEP study:

> spelt out the design failure of large modern estates and underlined the need for compensatory management if flatted estates were to work. Coupled with design aberrations, the decline of localised housing management and the concentration of desperate households within unpopular estates were both direct causes of disintegration. (Power 1985: 524)

The bad reputation of council estates resulted from a combination of many factors: poor design and poor management, as well as anti-social behaviour arising from the specific social problems of the families living there (Power 1984a; 1984b, 1985; interview Sheena Wilson, 2012).

The reputation-enhancing delivery of policy-influencing research by the DoE's SRD/HDD housing directorates throughout the 1980s was in contrast to the perception of the Department's Inner Cities Directorate. Despite the urban programmes that the Inner Cities Directorate had been running for several years, this team was belittled for its apparent lack of impact in the wake of the riots in

Brixton and Tottenham in 1983. The shock of the riots prompted a call for a more locally based approach to housing and regeneration policy. The PEP, recommending local estate-based housing offices and residents' involvement, seemed a tailor-made approach. Yet PEP was not felt to be progressing far or fast enough and in 1985 Michael Heseltine and Sir George Young (then Minister of State for Housing) set up the UHRU to target government intervention directly into poorly performing housing. The UHRU initially concentrated on urban housing authorities, with the remit to identify failing, unpopular estates and to devise a targeted funding programme to address their problems. Sir George Young found the title 'Urban Housing Renewal Unit' an unwieldy mouthful and the name of the team and programme was quickly changed to EAP.

The EAP ran from 1986 to 1994 (when it became part of the SRB) and had similar aims to PEP including involving the local community through consultation. EAP gave councils the opportunity to bid for resources for problem estates. It also began the slippery road of encouraging housing associations to become more involved, for it was difficult to transfer problem estates to housing associations through the new stock transfer programmes. The EAP was a substantial government programme that had spent £1,975 million by 1995 covering 540,000 homes (Towers 2000: 107). By 1987 the EAP remit was enlarged to include all housing authorities across England, rather than just 'problem' inner-city estates. Schemes were funded with money diverted from the existing Housing Investment Programme (HIP). This was not initially popular with council housing delivery teams who were also unhappy with the requirement to diversify tenure. The Treasury felt that pilots were needed for the first physical improvement grants, before any large-scale funds could be allocated. After much negotiation, an initial £15 million was released with agreement that further money would be top-sliced from the HIP programme. A similar process of top-slicing bids was later repeated to provide funding for the DICE program (interview John Harvey, 2009)[5] (see Figure 4.4).

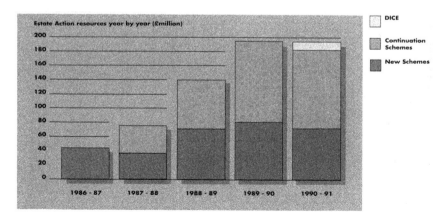

Figure 4.4   Top-slicing of EAP to fund DICE. (Source: *Estates Action Annual Report*, 1990–1991, DoE, p. 4)

The PEP was perceived to be a 'live experiment', to demonstrate the importance of effective localised management approaches in the regeneration of run-down estates. The essence of the PEP was establishing a local management office responsible for landlord's services and caretaking, whilst giving tenants the opportunity to exercise control over their homes and neighbourhood. Reduction of crime and vandalism were not the main focus of the PEP approach, yet these were seen as important goals.

Between 1987 and 1993 the HO conducted evaluations into the impact of PEP on all aspects of community life and crime.[6] Hope and Dowds (1987) developed a model (Figure 4.5) explaining how the PEP interventions increased residents' informal control over their estates, which might positively reduce crime and incivilities. Residents' collective commitment to the estate, arising from concern for its physical conditions and standards of social conduct, would be influenced by

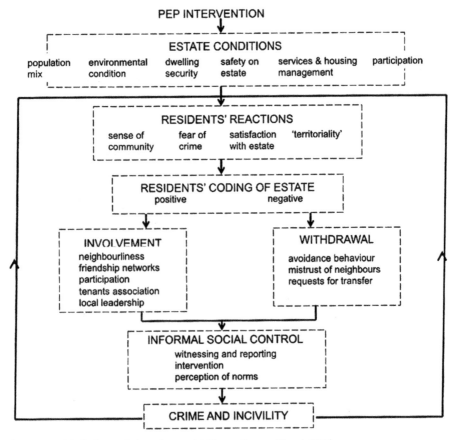

Figure 4.5   Priority Estates Project crime model. (Source: Hope and Dowds 1987)

their positive or negative 'coding' of the estate and individual sense of community, territoriality, fear of crime and satisfaction with the estate. The findings from the initial 20 PEP estates were highly positive with most of the estates improving rapidly. The evaluation studies by the DoE and Anne Power at the LSE, established a powerful and widely used body of best practice for delivering housing services and involving tenants in the day-to-day running of their estates such as the *PEP Guide to Local Housing Management* (Power 1987). By way of contrast, the research the DoE undertook into six EAP schemes concluded there had been limited effectiveness. Only one of the six showed real value for money, while for the others the results were more ambiguous (DoE 1996: 33).

Community Architecture and Defensible Space

One of Prince Charles' informal advisors (see previous chapter on his 'shifting kitchen cabinet', which included Alice Coleman) was the architect Charles Knevitt, who coined the term 'community architecture' to describe urban planning dictated and shaped by local residents rather than imposed by professionals. Knevitt was the architectural correspondent of the *Sunday Telegraph* (1980–1984) and then *The Times* (1984–1991). In his 1975 article in *Building Design*, which made public the term 'community architecture', Knevitt praised architect Rod Hackney for his work saving Macclesfield terraces from slum clearance and instead improving them in the way that their inhabitants wanted. Over a decade later, in 1987, Knevitt published a book with Nick Wates on the subject: *Community Architecture: How People are Creating their Own Environment* (Wates and Knevitt 1987). With Rod Hackney elected as RIBA president in 1987, community architecture became widely voiced as a solution to the issues of urban blight, endorsed by eminent figures such as Robert Runcie (then Archbishop of Canterbury) and the Prince of Wales. Knevitt and Wates put on a number of conferences on community architecture, including *Building Communities* at the Institute of Contemporary Arts in 1986, at which the Prince of Wales spoke, and later at the *Remaking the Cities* conference in Pittsburgh in 1988.

Prince Charles first lauded community architecture in his now infamous 1984 'monstrous carbuncle' speech at an event to celebrate the 150th anniversary of RIBA, in which he also repudiated modernist brutalist architecture:

> For far too long it seems to me, some planners and architects have consistently ignored the feelings and wishes of the mass of ordinary people in this country ... What I believe is important about community architecture is that it has shown 'ordinary' people that their views are worth having; that architects and planners do not necessarily have the monopoly of knowing best about taste, style and planning: that they need not be made to feel guilty or ignorant if their natural preference is for the more 'traditional' designs – for a small garden, for courtyards, arches and porches....

In his 1984 speech Prince Charles also implicitly supports defensible space ideas:

> Apart from anything else there is an assumption that if people have played a part in creating something they might conceivably treat it as their own possession and look after it, thus making an attempt at the problem of reducing vandalism.

This speech was seen as a 'watershed in helping to achieve the political break-through in community architecture' (Wates and Knevitt 1987: 20). The prince was said to have been introduced to the notion of community architecture by Jules Lubbock, architectural critic for the *New Statesman*, at a Royal Academy event, after which Lubbock sent him some clippings (Wates and Knevitt 1987). Another suggestion is that the ideas were introduced to Prince Charles somewhat earlier by the Duke of Gloucester, Prince Richard, who was a former partner of Bernard Hunt and John Thompson at HTA (discussed later). The three had read architecture at Cambridge together and founded their architecture practice in 1969.

Similarly to Alice Coleman, who saw her work as apolitical (Guest 1990: 20, interview Hilary Macaskill, 2013, see Chapter Three), Wates and Knevitt (1987: 21) argued that 'community architecture is not political in the party political sense of the word ...it transcends traditional Left/Right politics'. Jeremy Till (1998: 37) sees this political ambivalence as part of a wider drift towards a politics of consensus that ends up supporting the status quo. Of course, as Till (1998:37) points out, community architecture's most influential champion – Prince Charles – must remain apolitical; nevertheless, his voiced concerns at the time over no go areas and racial tensions were said to have angered Thatcher and the Queen. In 1987 Knevitt published a front-page splash in *The Times* on Prince Charles' 1987 speech to the Corporation of London in which he accused developers and architects of having done more damage than the Luftwaffe. Over a period of months Prince Charles visited a number of different community architecture projects across the country, one being the Lea View House estate in East London (the geography field trip mentioned in Chapter Three). This was a high point for community architecture.

One of the United Kingdom's most established community architecture firms was Hunt Thompson Associates (HTA) which pioneered the practice of community architecture combined with social research. Cofounder John Thompson's formative experience in the 1970s and 1980s was rooted in design practice in urban areas experiencing significant social and economic challenges. At a time when involving local people in the design of their living environments felt innovative and experimental, HTA undertook the remodelling of Lea View House in Stamford Hill, Hackney. Built in the 1930s Stamford Hill was one of the larger London County council estates. The 300 flats in Lea

View House were described as 'Heaven in Hackney' by their first occupants, however 40 years later it was crime-ridden, hard to heat and deeply unpopular. In 1980 90% of residents wanted to move away.[7] HTA had been in practice for a decade and were beginning to be commissioned to build new housing association homes in the late 1970s when they were first asked to visit Lea View House:

> We were absolutely horrified. We'd never really been on a public housing estate before. The place was disfigured by dripping overflow pipes from all the cisterns dripping down the faces of the building. It froze that winter and you got icicles hanging off the faces of the buildings. It was covered in graffiti. It was a five-storey walk up block, one big courtyard, open access staircases, the same flat on every floor. People cried when they were allocated a place on the estate. You couldn't get newspapers delivered. It had a terrible reputation and everyone more or less thought it was the people's problem, you know, it was a problem generated by the community living there. (interview John Thompson, 2013)

John Thompson's response was to move a project office onto the estate for two years. Initially facing a hostile reception (the project office was vandalised) he started with a social survey to understand the tenants and their needs better, including their attitudes to the place. This study data informed consultation events where residents developed the solutions they felt would work best. Through embedding themselves in the life of the estate, talking to people, they got to know the residents – 'tapped into local knowledge'. 'We turned the estate round'; so much, that by 1985 90% of residents reported they wanted to stay at Lea View House.

This dramatic change relied on a collaborative, hands-on approach to the design strategies and their application. The identical flats were converted into maisonettes for families, sheltered accommodation, and small groups of flats for people without children. Everyone could customise their kitchens and wallpapers, homes were fully insulated, with new windows and central heating for the first time. Solar thermal panels were installed for the south-facing maisonettes. For the first time, works were done on a rolling programme with residents in situ. So, the contractor worked on three staircases and as soon as the first one finished, the fourth staircase moved in, meaning most residents moved straight into a newly transformed home. Landscaped front gardens and secure entrances from each of the surrounding streets led to new lifts into the blocks with carpeted communal circulation. The hard courtyards were transformed by private gardens and children did the communal landscaping. John Thompson pointed out that the original scheme had 'exactly the same design as Trinity College, Cambridge. Open courtyard in the middle, staircases with people living in it, except they were families or different people, and it just didn't function' (interview John Thompson, 2013).

Describing Lea View House as 'a completely seminal moment', Thompson reflected on how the practice might have taken a different route:

We had six months to spend £6 million improving spaces without moving the people out and giving them what they said they wanted. We could have done that and we'd have made a fortune out of it. We spent the same amount of money much slower and it created a career. It was transformational in my life and it made HTA … but we could have missed it. We could have just put another team from the practice … there was nobody who particularly wanted to do it because it wasn't … well, it wasn't 'architecture'. (interview John Thompson, 2013)

Thompson had been hoping for a commission to build a new piece of architecture. If the existing residents had already been decanted and the place had been empty Hackney council may have ended up with a new block of better-quality homes, but not learnt from a pioneering resident-led approach to modernising existing estates, which demonstrated ways to resolve the long-standing community dissatisfaction:

We wouldn't have gone through the eye of a needle, and we wouldn't have had the conviction that it was the structural organisation … It was the placemaking within it. We realised that there was nothing wrong with the people, but there was just a tangle of people's lives, which made the place dysfunctional. That looked like a massive problem and, of course, solving that became ground breaking. (interview John Thompson, 2013)

The remodelling and renovation took four years and cost £6.5 million with the first homes occupied in July 1983. At Lea View House Thompson learnt that engaging with the community, and helping them articulate what they wanted from their homes, was an essential design exercise, and that their views were as important in finding lasting solutions as more technical inputs. In contrast, Coleman's DICE residents' engagement in design was a much more top-down and superficial process.

Doing DICE: Putting Defensible Space into Practice

As defensible space moved from being an object of study to practical technique, it is useful to frame the discussion through the idea of assemblage because it draws attention to the heterogeneity of the elements that come together in the 'doing': people (agents), objects (programmes, documents, drawings, plans, etc). Using the notion of 'assemblage' allows us to be open to the possibility of ongoing dialogue between and across the research sites (estates), ideas on defensible space and the key agents in its mobilisation.

Having acquired the support of the Prime Minister and the promise of extensive funding for her DICE project, Coleman immediately began to establish a DICE research unit in the Geography Department at KCL. An unusual hybrid mix of research and practical architectural skills were required for such a complex study. Coleman had funding for five years, an extremely short period of time for a single major regeneration scheme, let alone for her initial target of nine estates. But the sense of urgency, which had flavoured Coleman's conversation with Margaret Thatcher in January 1988, dissipated amongst the machinery of departmental bureaucracy and it took 21 months before the DoE officially started the project in November 1989 (Price Waterhouse 1997a).

Initially the DICE team consisted of Coleman, a couple of KCL geography researchers, three postgraduate students, a housing consultant, and then later architect Mary McKeown. At the outset the team was conspicuously light on construction experience and, while Coleman was confident that she could provide planning guidance, architectural skills were needed. Mary McKeown was working as an architect in the maintenance department of the Northern Ireland Housing Executive. There, she became familiar with government research on reducing crime in housing through her work refurbishing bomb-damaged estates. Coleman visited Belfast in the Autumn of 1985, undertaking a DICE survey of the infamous Divis Flats. McKeown attended a public lecture given by Coleman in 1986 (Bar-Hillel 1986b), this prompted her to read *Utopia on Trial* which she found 'put into rational form much of what we felt was wrong and needed to be fixed' (interview Mary McKeown, 2013). McKeown was appointed to the DICE unit, transferring from a practical sphere to a research one:

> I'd left a desk where I was dealing with tenants and damp, and arrived in a hornets' nest of academia. (interview Mary McKeown, 2013)

Architect Peter Silver, who replaced McKeown in 1991, came from a similarly practical housing background. Having worked for the housing cooperative Solon, he graduated from the Architectural Association, before being employed as a research assistant on DICE.

Most of the DICE research grant money went to capital regeneration costs on seven DICE estates: Rogers and Ranwell East in Tower Hamlets; Kingsthorpe Close, Nottingham; Avenham Estate, Preston; Bennett Street, Manchester; Nazareth in Birmingham; and the Durham Estate in Sandwell (see Figure 4.6 on all the council estates Coleman worked in over her career).

Coleman's positivist epistemological framing of the research meant that having formulated the hypothesis set out in *Utopia on Trial*, the next step in her scientific experiment was careful selection of the objects of study (see Figure 4.7). This selection would shape the objectiveness of her research evaluation (Palfrey et al., 2012). Choosing the DICE estates, however, was a drawn-out process: inviting Local Authorities to submit potential estates to the DoE,

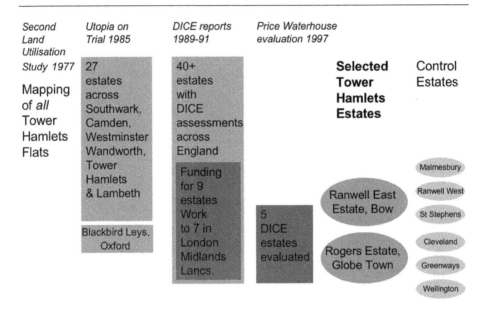

Figure 4.6   Coleman's surveyed estates.

followed by an initial assessment survey, leading to design proposals, culminating in a consultative referendum with estate residents, before making a decision on the estate's inclusion in the study. Each of these steps required Coleman's strict academic experimental model to engage with Schön's (1983) messy 'swampy lowlands' of real-world problems: the competitive scramble for capital funding, the exploratory creativity needed for feasibility designs and, not least, the frustrations of residents engagement.

Over 40 estates were surveyed with 33 *DICE Disadvantagement Reports* written in an intensive thirty-five-month period between early 1989 and November 1991. Because of the compressed timeframe, the small KCL-based DICE research team was concurrently visiting potential sites, undertaking surveys and data collection, holding 'endless' meetings with residents, architects and Local Authority representatives, alongside drafting the reports. There was an uneven geographical spread of sites. Half of the locations were in Greater London. Outside London, Birmingham and Manchester each proposed two or three potential sites, with individual estates scattered across the Midlands and further north across Tyneside. At least five of the London estates evaluated were in Tower Hamlets (noticeably the largest cluster), and four estates measured in *Utopia on Trial* were revisited, including the Mozart Estate in Westminster (see our case study that follows Chapter Four).

Figure 4.7   Cartoon of DICE as scientific. (Permission: Simon Harding). (Source: *High Rise Quarterly*, 1989, 3: 3: 5)

Data collection visits were intense, rapid events, with the researchers spending a single day walking the estates. There was a rigorous process for completing the 'design-and-abuse' survey. Using a pre-coded proforma each block or house on the estate was assessed. The presence of litter scored one negative point if it was 'Clean or Casual' and two if it was 'Dirty and Decayed'. Following Coleman's geographical mapping background, site plans were marked-up with pre-selected colouring pencils classifying the space according to its privacy or public accessibility. From the completed survey sheets a 'before' design disadvantage score was calculated for the estate. Design changes were proposed and the design disadvantage score recalculated. A *DICE Disadvantagement Report* was drafted to brief consultants and as a resident consultation document. The standard report template consisted of an account of the DICE design principles, analysis of the estate design, the abuse surveys, and colour-coded before and after site plans.

This report formed the non-negotiable brief for the scheme although the KCL DICE Unit did not write a specification or cost the alterations. The design improvements were presented to Local Authority councillors and staff at tenants' meetings where residents voted on whether to accept the proposals. If a majority agreed to the scheme, Coleman would recommend their estate to the DoE. The scheme would then be submitted to the Local Authority planning committee and, if approved, back to the DoE for agreement to proceed to the working drawings stage. Gaining the cooperation of the statutory authorities was as great a challenge as gaining the acceptance of residents:

> Planning officers were going to have to sanction increased densities, fire officers were going to have to weigh the reduced likelihood of arson against a relaxation in the means of escape, housing departments were going to have to produce complex decanting programmes and often provide temporary re-housing; it was after all essential for the experiment that the original residents either remained or returned at the completion of the works. (Silver 1995: 6)

By the end of 1992 the DICE Unit had produced a comprehensive survey manual, with classification conventions for spaces similar to the instructions for the Second Land Use Survey which Coleman had masterminded[8] (Silver 1995). Yet McKeown's impression was that potential suitability was assimilated by one short walk across the estate. The decisive criterion for site selection was Coleman's perception of the extent that the design disadvantagement score could be reduced:

> Alice's target was to reduce the DICE defects from say 14 to 5. Alice would walk around an estate and guess the score might be a 13 but we can only get it down to a 10 so we won't do it. (interview Mary McKeown, 2013)

This was not an exact process (science) but other subjective factors influenced the decision such as Coleman's view on whether she could work with the Local Authority or if residents were prepared to be 'DICE-ed'. The refurbishment opportunity was usually presented as a stark take-it-or-leave it offer:

> The residents' meetings I attended, were not like residents' meetings I used to have at Solon, where everyone's voice had to be heard and everyone had an opinion. DICE were; 'Here's some information. This is our research. If you say yes, then you go with it. You can say no. You don't have to be DICE-ed'. (interview Peter Silver, 2013[9])

Contrasted with the methods of community architects, Coleman's inflexible theoretical approach inevitably overrode tenants wishes (see Figure 4.8):

> There was nearly always a fight between Alice and the residents, about what they wanted and what she was prepared to give them. If it didn't fit the theory she

wouldn't allow them that. From an experimental point of view, she was right. If you're going to test a theory, then test just the theory. (interview Paul Wiles, 2011[10])

There was also a degree of compulsion exerted on residents by Local Authorities keen to receive the money. This was despite authorities having to commit to spend any additional funds needed for repairs, installation of communal heating systems, pitched roofs or new street lighting, which were not eligible under the DoE funding. In the main residents were keen to agree to works that would improve their homes and surroundings, but there were a number of meetings where residents were unenthusiastic or had to be cajoled into agreement. At the Rogers Estate 74% of residents voted for the works. Yet the chairman of the Rogers' residents' association did not vote for the DICE proposals, as he believed there had been insufficient tenant involvement in the designs (Guest 1990). Ranwell West and Ranwell Close (1938 courtyard blocks on the northern edge of Ranwell East) both voted not to be included (interview Michelle Smith, 2011[11]). The proportion of tenants voting in favour of the works at Ranwell East was high

The 'DICE' project at King's College (page 4).

Figure 4.8   Cartoon of Alice Coleman's DICE and the treatment of tenants. (Permission: Simon Harding). (Source: *High Rise Quarterly*, 1989, 3:3:1)

at 86%, however the scheme was highly unpopular with a vocal minority of the residents, which Coleman dubbed 'aggressive opposition from a few militants' (Silver 1995: 3). There were also suggestions of a political tint to those selected:

> What was wrong about DICE was she [Coleman] was poacher and gamekeeper, she set the rules and the Local Authority couldn't say anything. If they did she wouldn't work with them. Conservatives wouldn't work with her, Lib Dem councils might, but Alice wouldn't work with Labour councils. (interview Mary McKeown, 2013)

Earlier DoE regeneration programmes such as PEP had already targeted estates with the poorest conditions. Neighbourhood Manager, Steve Stride remembered that both the Rogers and Ranwell East 'weren't the worst in the borough', but both 'had the signs of serious dysfunctionalism' that could have eventually led to their demolition (interview Steve Stride, 2011). Ranwell East had already received several rounds of regeneration investment and had been the site of two SNU improvement projects between 1981 and 1986. Compared to the highly problematic estates surrounding it, such as Lefevre Walk, Ranwell East was at the time seen as being in an acceptable state for an 'upcoming area' (interview Michelle Smith, 2011). This echoed the perception that selected DICE estates were not those with the worst conditions, or most extreme deprivation, but with the greatest potential to change (interview David Riley and Barry Stanford, 2011; interview Mary McKeown and Peter Silver, 2013).

Tower Hamlets as a Site of Innovation: Rogers and Ranwell East Estates

While each of the seven DICE estates would have been worthwhile to examine, we have examined just two in any detail – Rogers and Ranwell East Estates in Tower Hamlets. These two estates were selected because we only had the time and resources to study two closely (including undertaking interviews with a handful of long-time residents) and as they demonstrated maximum variation across several physical and social variables. They were very different in scale, built form, density, severity of the social/crime problems experienced and the design solutions proposed. It was only on completing the detailed investigation that a distinct disparity in the success of the projects and the attitudes of the individuals involved emerged.

By the 1980s, Tower Hamlets and Southwark were the two London Boroughs that contained the most council flats (Coleman 1985a). In *Utopia on Trial*, the mean design disadvantagement score for all the Tower Hamlets homes was assessed at 8.3, slightly greater than those in Southwark, but Tower Hamlets had a substantially worse 'abuse' score. At that time Tower Hamlets in inner East London had a reputation for housing poor and underprivileged communities who subsequently ended up living in close proximity to the high-income wealth generating employment areas of The City and Docklands. Despite improvements in poverty over time, today Tower Hamlets is ranked as the 10th most deprived Local Authority in England, and areas of deeply embedded poverty remain. At the time of DICE many of the estates had not been refurbished since their construction, with the Council lacking resources to improve the neglected housing stock. Nonetheless, following the Liberal Democrats taking control of the Council in 1986, Tower Hamlets became notable for applying innovative forms of neighbourhood management. In a precursor to the current localism agenda (Localism Act, 2011), they established seven neighbourhood management areas each delivering a decentralised housing service from a local Neighbourhood Office. This culture, fostering fresh approaches to housing management, attracted innovative-minded staff. One of these was Steve Stride, the Neighbourhood Manager for the Globe Town Neighbourhood in Bethnal Green.

Rogers Estate

Constructed in 1949, Rogers Estate was the smallest and oldest of the seven DICE schemes. 124 flats were split between two well-defined five-storey U-shaped deck-access blocks.[12] Entrances to communal stairs were dark, doorless and tucked away out of sight. The site was open, with large ungated grassed areas leading uninterrupted from the street up to the windows of ground floor flats. In the bid submitted to the DoE, the estate was unsurprisingly painted in unflattering terms, portraying it as deserving of investment:

> The Estate suffers from a number of endemic problems common to many inner-city environments, namely litter, waste tipping, vandalism, graffiti and crime. All of these worse than average for the locality and all contributing to an atmosphere of environmental degradation and social malaise. (Price Waterhouse 1995: 19)

While the problems listed are typical, the use of 'social malaise' as a phrase is telling, as it strongly echoes Coleman's terminology in *Utopia on Trial*. The project manager at Rogers, Sam McCarthy (see Figure 4.9), softened this description, attributing the general air of tired neglect to a widely prevalent lack of maintenance. Rogers was in a bad, but not unusual, condition:

> It was typical of a number of estates within Globe Town. There'd been a lack of expenditure on it. The amount of funding that the Council and the neighbourhoods had just wasn't enough to sustain the buildings. And also you get used to a design and think it works and then you don't realise it's not working until something like DICE comes along. You know, rat runs, no defensible space, cut-throughs on the estate. It was in a very rundown condition and in need of attention. (interview Sam McCarthy, 2011)

When surveyed, the pre-DICE design and abuse scores for Rogers were far higher than the borough average, with the average design disadvantagement score for the Rogers Estate at 11.5 and the abuse score measured as 12:

> Such scores inevitably spell social breakdown, so DICE proposed changes to remodel the defects in ways that would render them harmless. The final average [following remedial work] would have been 1.5, but a last-minute fiat by councillors raised it to 2.4. (Silver 1995: 4)

The redesign of the Rogers Estate (see Figure 4.10) altered the scale of the site, remodelling communal areas and individual homes. A new road cut the site in half, creating two distinct 'island sites', redirecting pedestrians, particularly school children, around the estate. Ground-floor flats were reorientated with new front doors opening onto front gardens with waist-height brick walls facing the street. Individual back gardens replaced courtyard car parking. A terrace of eight new bungalows were built, completing the street frontage onto Globe Road. Parking was kept in the smaller South-block courtyard, but ground-floor flats were given front gardens as buffer zones. A long-term North block resident

Figure 4.9   Sam McCarthy on the Rogers Estate. (Source: Building Magazine 1990, 255, 20-22)

whose ground-floor flat overlooked Sceptre Road recalled the earlier use patterns through the courtyard:

> You could drive in one side and out the other. Kids played in the front [Globe Road]. There was no play space but kids played there, everyone was out there. (interview long term Rogers Estate resident, 2011)

This resident still associated the old 'front space' as a communal shared space and was dissatisfied with the experience of living facing a street, complaining that before the 'flats turned back-to-front' he used to know everyone, but not now. This is telling, as Coleman's aim was to reduce anonymity and increase the sense of community by reducing the occupation of shared spaces. Less people circulating on each corridor and entrance would facilitate better recognition of intruders. Blocks were split internally into smaller sub-blocks, cutting access balconies to serve only one or two dwellings. Nine new entrances and four new lift towers were built with well-lit, glazed entrance doors. All entrances faced onto roads to encourage surveillance. In addition to the DICE works the Local Authority took the opportunity to carry out other improvements, which were not critical to the DICE experiment: these consisted of adding pitched roofs, installing double-glazing and new heating systems, as well as replacement kitchens and bathrooms. Initially the 1991 DICE contract costs for the Rogers Estate were £3.5 m, and despite increasing to £4.7 m by the contract close, the works were considered to be very good value by the project team.

The newly completed Rogers Estate, visibly improved and easily accessible from Westminster, was a powerful good practice case study; a showcase for urban regeneration and a transfer mechanism for spreading the concept of defensible space:

> George Young[13] loved it. He came down and brought hundreds of civil servants down, loads of times … We had coach loads. He loved Bow Town. He saw us as a model and loved us. (interview Steve Stride, 2011)

Whether Coleman's ambitious predicted outcomes for the remodelled estates were achieved is discussed in Chapter Five referring to the Price Waterhouse evaluation that explored the mass of data collected at the time, measuring decreased estate management costs, resident turnover and reported crime. Under Coleman's key indicators of design disadvantagement abuse scores, her DICE interventions predicted a decrease for the Rogers Estate from 11.6 to 1.8. Yet the positive impact on vandalism, graffiti and litter was not sustained. Rogers, which had a measured abuse score of 12 pre-DICE, dropped to 2.4 post-DICE but rose again to 4.2 when resurveyed a year later. Coleman blamed the worsening abuse scores on 'other forms of social breakdown' and on the shallow forecourts that collected rubbish, ignoring alternative explanations, such as the unpredictability

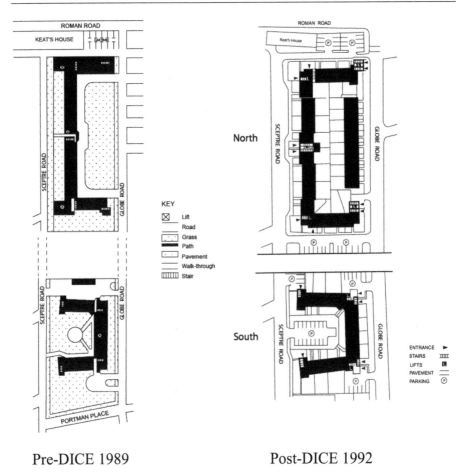

Pre-DICE 1989   Post-DICE 1992

Figure 4.10   Rogers Estate site plans pre- and post-DICE. (Source: Evaluation of Design Improvement Controlled Experiment (DICE); Report on Rogers Estate, Globe Town 1995)

of litter as a measure, particularly for a site with a corner shop on the route to a nearby school.

The Price Waterhouse evaluation attributed the reduction of crime levels on the Rogers Estate after the DICE project to the general gentrification of the wider area rather than the works themselves. As a controlled experiment, both Coleman and Price Waterhouse attempted to isolate and measure the impact of a single initiative amongst a range of complex interconnecting changes (Harrison 2000); whereas the Globe Town Neighbourhood team saw the works to the Rogers Estate as a small part of the jigsaw of improvements made by their intense local management of the neighbourhood. Revisiting the Rogers Estate in 2011, the DICE project manager Sam McCarthy's professional assessment of the long-term upkeep was less positive, recognising lack of funding as restricting maintenance to a less than reasonable standard (interview Sam McCarthy, 2011). He spoke to an original resident of the bungalows, reporting 'she still loves it, she absolutely loves it'. However, another resident bitterly recalled two decades later, 'the only thing I got from the works was a back garden to mow' (interview long term Rogers Estate resident, 2011), implying that any benefits had been outweighed by the fuss of the redevelopment.

Ranwell East Estate

The nearby Ranwell East Estate was an architecturally distinct but rambling estate, separated from the surrounding street network; it sprawled over an area of 5.75 hectares with internal roads, paths and green spaces (see Figure 4.11). Constructed in 1974 it provided 474 homes within 21 four- and six-storey slab blocks, 14 of which were linked by overhead walkways. The blocks had maze-like internal corridors interconnecting all the homes. Any entrance could lead to all other exits via the walkways and lift/stair towers. The enclosed spine corridors were dark and unobserved by either homes or outside spaces. Six of the blocks were designated for elderly residents, yet the overall provision was mainly single-bed flats and high-level maisonettes inappropriate for Tower Hamlets' housing policy, which, at the time, aimed to allocate houses to families.

Figure 4.11   Ranwell East Estate pre-DICE. (Source: Evaluation of Design Improvement Controlled Experiment (DICE); Report on Ranwell East Estate, Bow, 1996)

A council employee, Michelle Smith began her career as administrator for the Ranwell project team, eventually returning to the estate as housing manager. Smith remembers the local area as rundown and the Ranwell East Estate pre-DICE as a harsh 'concrete jungle'. The public spaces were bland, unfenced expanses of grass with little planting. One central space contained a children's play area above the roof of a large underground carpark. There were frequent complaints that the communal refuse bins were too small and overflowing with rubbish.[14]

The Ranwell East Estate blocks were joined by link bridges at first, second and fourth floor levels (see Figures 4.12 and 4.13). Smith identified these as areas where groups of youths congregated, groups who were reported as causing disturbances and muggings (interview Michelle Smith, 2011). Coleman identified this as Ranwell East's worst defect:

> The complex was a maze of escape routes for hooligans and criminals and because anyone could go through all other blocks, tenants could not get to know everyone frequenting their territory and a spirit of anonymity and alienation prevailed. (Silver 1995)

Coleman had previously surveyed the estate for *Utopia on Trial*, noting its notoriety as a failing estate. Her assessment visit in 1988 identified an average design

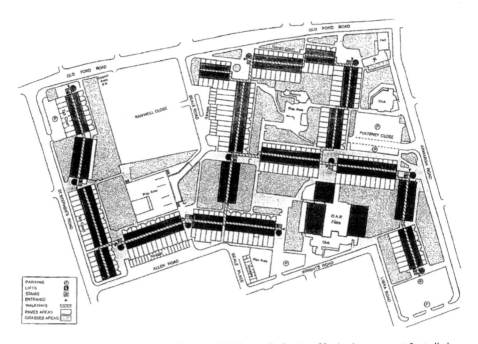

Figure 4.12   Ranwell East Estate site plan pre-DICE. (Source: Evaluation of Design Improvement Controlled Experiment (DICE); Report on Ranwell East Estate, Bow, 1996)

Figure 4.13   The walkways, Ranwell East Estate. (Source: Inside Housing Magazine 24 May 1991)

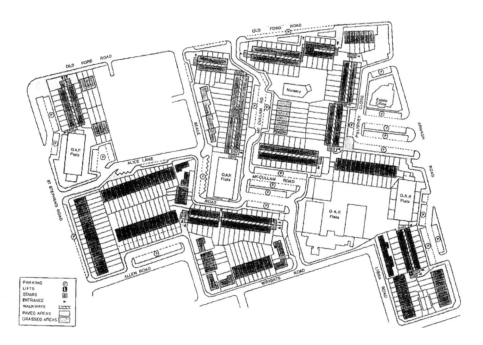

Figure 4.14   Ranwell East Estate post-DICE. (Source: Evaluation of Design Improvement Controlled Experiment (DICE); Report on Ranwell East Estate, Bow, 1996)

disadvantagement score of 11.9 placing it amongst the worst 10% surveyed (Coleman and Naylor 1993). Substantial change would be needed to reduce this score and the DICE Unit suggested extensive alterations (see Figure 4.14). The scale and complexity of these works resulted in the highest cost per unit of all seven DICE schemes. Costing over £18 m in 1991, the Ranwell East Estate works were the largest estate improvement contract in the country at that time (Silver 1995). It was also a sensitive selection, having already received substantial improvement and crime reduction grants.

Pre-DICE the Ranwell East Estate was no stranger to other attempts to improve it, indeed it was 1 of 18 estates where the SNU ran practical crime reduction projects between 1981 and 1986. Their 1988 report recorded two design problems: a significant lack of security to individual blocks and poor heating resulting in very high bills. Both were accentuated by inadequate cleaning and an unresponsive repair service. SNU identified two main crime and anti-social behaviour issues; vandalism and noise nuisance. They recorded high child densities and suggested that anti-social behaviour was aggravated by the lack of social and play facilities. The number of children on both estates exceeded Sheena Wilson's (1980) recommended threshold for child density, which she linked to high levels of vandalism. Both estates also had large numbers of elderly residents and it was this mix of young and old that exaggerated the generational conflict. Youths congregating was a frequent complaint on both the Rogers and Ranwell East Estates. These population characteristics will have affected the application of defensible space principles, such as density of occupation, the presence of residents during the day to provide 'natural surveillance', or whether someone was likely to intervene if they see criminal or anti-social behaviour. The SNU was disappointed with the lack of sustained impact from their own interventions at Ranwell East, which consisted of deploying youth workers and small-scale security improvements: 'despite being one of the Unit's earliest projects, it is the one in which least progress has been made' (SNU 1988a: 25).

The DICE redesign of the Ranwell East Estate demolished the walkways and three link blocks so no interconnecting entrances remained (see Figure 4.14). External balconies were installed at the first and third floor levels of the larger blocks, doubling as a secondary means of escape for upper floors (Price Waterhouse 1996). Sub-dividing the blocks to meet Coleman's ideal threshold of six homes per corridor was challenging. New stair towers broke up the four-storey blocks into two stacked maisonettes. On higher blocks only one lift per block was affordable, having to serve a whole upper floor. Some internal modernisation was undertaken in parallel with DICE, with replacement central heating systems, new kitchens and bathrooms, and UPVC windows installed. The gaps left by the demolished slab blocks and underused garages were infilled with 108 new, traditional terraced houses. Pre- and post-DICE the overall number of homes remained similar, but funding the new housing proved problematic. DICE funding paid for 234 new-build houses across the seven schemes, but there was a waning of DoE support for new construction, which Coleman attributed to the loss of Thatcher's personal support after her resignation (Silver 1995). But in fact Thatcher's housing policy was one of the barriers to the construction of new housing. By the mid-1990s Councils were unable to retain funds received from right-to-buy and were:

> saddled with a housing policy which didn't allow them to build houses. Alice just couldn't understand, building a house is very different to putting up a fence. (interview Steve Stride, 2011)

New housing could be paid for by HIP but was to be run and managed by a housing association such as the Old Ford Housing Association, which later took on ownership of the Ranwell East Estate.

Site layout changes and winding new roads aimed to channel traffic and pedestrians in front of homes (see Figure 4.15). The new houses, meandering roads and individual gardens created from previously public space blocked former pedestrian desire lines. Reworking the estate's layout by reorientating flats so front doors overlooked the new roads and inserting new entrance stair towers

Figure 4.15   The construction of Alice Lane, named after geographer Alice Coleman. (Source: Peter Silver)

resulted in extremely complicated and confusing house numbering. One block had three road addresses, with non-sequentially numbered flats.

However, the Ranwell East scheme was affected by more serious delivery and contractual problems. In 1994, towards the end of the remodelling, *The Observer* reported an investigation by Scotland Yard's Fraud Squad into accusations of maladministration of a tender for design work in Tower Hamlets. This professional impropriety resulted in the removal of the contract manager. The improperly selected architects' contract had been terminated earlier when Coleman saw their proposed designs. In her opinion 'they weren't up to the job, that was obvious' (Dodd 1994: 6) resulting in the scheme being designed by temporary architecture staff employed by the council. This incident may seem an entertaining but irrelevant 'local government waste of public funds' scandal, but it is also an example of Coleman's significant influence in achieving the design consultants' dismissal from the job, not for improper tendering, but for their failure to implement her defensible space principles. It is a reminder of the dangers of blurring academic research into commercial/professional practice particularly in highly procedural activities such as public sector procurement. It is not clear whether the poor original architects or the change of staff can explain the arbitrary and off-kilter design decisions visible on the estate. Additional security features (such as door entry phones originally vetoed by Coleman, or metal screens put on to limit access and preserve privacy) were rapidly retrofitted to rectify the negative consequences of the design changes. Coleman disliked door entry phones, which she described as 'a siege exercise, accepting the criminal presence and trying to keep it out' (Bar-Hillel 1986a: 17), arguing that suitably designed entrances were a sufficient deterrent to make them irrelevant. She criticised the HO's 'designing-out-crime' through security improvements, like locks and entry phones, as 'first measures, not fundamental cures' (Coleman 1989: 4). At Ranwell East, she had argued for the removal of the existing door entry phones, but in practice:

> Unfortunately, that didn't work, because no sooner did the new stairwells go up with the doors removed, than youths on the estate would congregate on the stairwells once again, which is what obviously the project wanted to get rid of. So the door entry systems did go back. (interview Michelle Smith, 2011)

A *Building* magazine article described the removal of the existing entry phones[15] as illogical, noting that the Council replaced them two years after DICE was completed (Spring 1997). A similar situation arose at the Rogers Estate, where the adaptations resulted in 16 entrances, a high number for two small housing blocks. Again Coleman vetoed entry phones, but following pressure from residents, post-DICE they were installed, with a resultant disproportionate maintenance burden (Price Waterhouse 1995). Despite reinstallation of entry controls in response to tenants' demands, the current housing managers interviewed were sympathetic to the significance Coleman placed on symbolic over physical thresholds. They recognised the trade-off between inherited ongoing maintenance costs for access controls (which are now seen as a normal basic specification), against Coleman's free but possibly less-effectual design-led solution (interview Michelle Smith, 2011; interview Kobir Choudrey, 2011).

In terms of informal dissemination mechanisms, this kind of incident or post-scheme modification is often gossiped about by architects, quantity surveyors or Local Authority housing staff. So it is likely to have contributed to shaping the narrative of Ranwell East as an [un]successful regeneration case study. Nevertheless, at Ranwell East, Coleman's design disadvantagement score decreased from 11.8 to 2.2, and it experienced a reduction in vandalism, graffiti, and litter too, from the 6.1 reported in *Utopia on Trial* (a score echoed in the 6.4 measured in the pre-DICE survey) to 4.1 post-DICE. Ranwell East occupants themselves valued gaining private external space:

> The best thing about DICE was they closed the landings and we got an extra bit of garden. (interview long-term Ranwell East resident, 2011)

Even when residents noticed the reduced public space, they were not critical of this change:

> There's less open public space. It is much denser but to be honest there's less vandalism and less congregating. (interview long-term Ranwell East resident, 2011)

At the time professional opinions on the outcomes at Ranwell East were varied. Some were not particularly favourable: 'Ranwell East's tawdry, 1980s spec-built image will date as quickly as did the estate's original 1960s vintage housing' (Guest 1993; Spring 1997: 50). Regardless of this scorn, three years after completion of the scheme, interviewed for *Building*, Michelle Smith reported:

> It's definitely a nicer place to live. Before the whole layout and feel of the estate was dull and dingy and people were moving out. Now, groups of residents sit in their gardens or on their balconies in summer and have drinks together. There's a community spirit slowly growing up. (Spring 1997: 50)

The operationalisation – putting into practice – of defensible space principles on these two Tower Hamlet's council estates coincided with an increased interest in tenant consultation in the 1980s into the 1990s. Levels of consultation appear to have varied between the two estates. A great deal of tenant consultation was undertaken on the Rogers Estate with door-to-door surveys as well as open forum meetings:

> It was going to be a big change for the residents. This wasn't just doing new windows and a bit of landscaping, we were changing the whole of the design, the layout

of the blocks. We were splitting friends up ... the old makeup, the dynamics of the block were being changed ... Many of them understood the principles behind what was trying to be achieved, because they'd been screaming and shouting formerly about the problems that they were encountering and we were saying we *can* alleviate these problems. (interview Steve Stride, 2011)

Sam McCarthy recalled residents' concern that the greater privacy would turn into isolation. Residents believed that shortening balconies to access only two to three flats would mean being cut off from their neighbours (interview Sam Mc-Carthy, 2011). This anecdote of a tenant having to exit the block and re-enter to visit an elderly neighbour in an adjacent flat was repeated in the mainstream press (*The Times* and *Daily Express*), as well as architectural publications (Guest 1993; Kanivk 1991; Warman 1991).

Rogers' residents were also worried about being moved away, but as with most DICE estates there was no decanting during the remodelling. Coleman's ambition for DICE as a controlled experiment required the resident population to remain unchanged, but the lack of community dispersal was as likely to have resulted from improved housing practices and the unpopularity (then) of large-scale decanting programmes with Local Authorities due to the costs and disruption.

Ranwell East residents were involved during the works through a steering group made up of leaseholders and residents meeting with the contractors every fortnight, but residents' memories show they were less than content with the consultation process and outcomes:

There were leaflets saying what was going to happen and a woman who did consultation. But they didn't consider what we'd already done. We'd paved the garden and the scaffolding cracked them ... We'd already done the brick wall and they put in another wall. It was all very stressful. M. used to go to meetings but it was typical, loads of people shouting, mayhem. You'd have to go to have your say even if they didn't listen. Some people were pleased with the result, those who'd moved from the tower into the houses. M. used to talk with the contract manager. He'd say yes to anything to keep you quiet. (interview Ranwell East resident, 2011)

We see in these two example estates how defensible space was operationalised through DICE, but we would question the degree of replicability and if there were lessons that could be used elsewhere?

It was very, very difficult to secure money of that ilk, three or five million pounds worth of money, to do Estates Action programmes. Our capital programme up to that time was pretty low-level stuff, very much scratching the surface of estate renewal. A new roof, windows, doors but nothing that was able to transform a complete estate. DICE was definitely a one-off opportunity to be honest. Something we wanted to migrate to other estates but that wasn't going to be possible. That doesn't mean we didn't take the ideology behind Rogers with us. (interview Steve Stride, 2011)

Many of the examples discussed in this chapter reinforce the impossibility of testing an urban policy with any experimental rigour. DICE was not operating in a policy vacuum and the project's aims were influenced by, and interacted with, other housing or economic policies of the time, as well as the urban fabric it set out to improve. Some constraints were intangible (tendering policies affecting management or repairs services, or the policies funding new housing), others physical (structural defects or existing road layouts). The recurring example of Coleman's rejection of entry phones as diluting the symbolic sense of owner-ship illustrates the conflict between her abstract theoretical position, residents' wishes and the practical experience of housing staff.[16] To practitioners such as the DICE architect Mary McKeown, with experience in the management and maintenance of large housing estates, the DICE principles failed to get to the causes of the physical deterioration on the estates, let alone beginning to address the social problems. She questioned the ethics of not addressing basic refurbish-ment issues:

> I looked at the DICE estates with windows falling out and wondered what we were doing just changing the outside spaces. I had a maintenance background, used to surveying for dampness and rot. I could see that the problems weren't going to be solved with some fences. (interview with Mary McKeown, 2013)

And occasionally practical experience won out, as the practitioners rejected Coleman's ideas regardless of the evidence she presented.

Many outcomes from the DICE project were non-monetary: for example, one result of DICE on the Rogers Estate was high profile attention for the wider Globe Town Neighbourhood management approach, the establishment of an experienced local team, and the gathering of knowledge and learning about defensible space. The Globe Town team used the DICE project and the publicity it attracted to justify increased levels of Estates Action funding for subsequent refurbishment projects. The team described DICE as an influential period in their lives and that this unique experience had supplied insights and practical lessons that they continued to apply. Still working in project management, Sam McCarthy described the experience of the DICE project as 'totally changing my thinking' around defensible space ideas:

> They're taken for granted now, and our design briefs automatically take them as first principles. (interview Sam McCarthy, 2011)

Indeed, McCarthy was so confident of the transferability of the DICE principles and their positive impact, that by mid-1993 he was planning to implement the best aspects, spending £6.5 m of Estates Action funding to apply DICE principles to the nearby Bethnal Green Estate. He ranked creating private front gardens as the simplest, most widespread, and successfully repeated feature. As well as achieving Coleman's improved territoriality and neighbourliness, it was inexpensive to implement, whist reducing Local Authority costs of upkeep for communal grassed areas.

Indeed, there were three other DICE legacy projects in Tower Hamlets: the Bancroft, Leopold, and Burdett Estates in Poplar. The Burdett was most closely a DICE scheme, where the whole estate was broken up, but at a cost of £55,000 a unit at 1998 prices (over £100,000 a unit today) it was very expensive. The DICE elements that could easily be replicated tended to be the cheaper ones: erecting fences, better lighting of public areas, only providing additional stairs to upper floors where this could be done cost effectively. These were mixed with familiar ideas like 'breaking up the box' or reducing deck access. The design proposals mutated as they adjusted to fit the new sites, cross-fertilising with other improvements such as removing refuse chutes that were additional to the DICE principles but that the team wanted to apply. When asked which aspects of DICE they would reapply now, the current Ranwell East Estate housing manager Robert Williams described a 'shopping list of ideas to try':

> We'd do bits and pieces. One of the things we've recognised quite quickly is there is not one solution. What we mean by defensible space will vary from area to area. Our recognition is that it has to be treated as a block by block, and an estate by estate type solution. (interview Robert Williams, 2011)

Williams identified elements not to repeat, particularly automatically installing high-level fencing as 'turning a place into Colditz'. Splitting communal grassed spaces into private gardens had resulted in unsightly poorly maintained gardens. Coleman's urge to build additional houses on every parcel of public green space, increasing site densities, limited the future adaptability of the estates; and once communal space has been transferred over to private ownership there was/is little chance of reversal.

There was no established delivery team at Ranwell East. The designs seemed contrived, appearing to have applied the DICE rules rigidly and unquestioningly – with insufficient sensitivity to the variety of flat types and block sizes on the estate. Yet the regeneration of the Mozart Estate (described in the case study following Chapter 4), comparable in size to the Ranwell East Estate, used an equally broad palate of responses, but each adaptation varied slightly over each phase and was adjusted depending on the response to previous adaptations.

In this chapter we have engaged with both making the operationalisation of defensible space principles more visible and looking at the practical issues involved in that operationalisation. What is clear is that in every day practice, objective 'evidence' has less persuasive status than a 'best-practice' case study or social mechanisms (e.g. the credibility of the individual passing on the concept, the recognition or perceived utility of the idea) to transfer and communicate 'what works'. As a result, by 1992 Coleman's design improvements were seen by many as a success story:

> Design improvement is important. It can make all the difference as to whether our country continues to plunge into moral decline and criminal activity; or whether people are enabled to climb back into a more civilised life. It is not just another gimmick, but based on hard factual evidence, derived from lengthy, painstaking research. (*Voluntary Housing* 1992: 13)

There were visits to DICE estates to 'see for yourselves how well it works in practice' (*Voluntary Housing* 1992), DICE presentations and even a design vetting service. Yet evaluation of DICE told a more mixed story, as the next chapter shows.

Notes

1 *The Psychology of Housing*. http://www.singleaspect.org.uk/?p=2363.
2 We deploy the notion of assemblage epistemologically (see DeLanda, 2016: chs 1–4).
3 *Building Safety Programme*. https://www.gov.uk/government/publications/building-safety-programme-monthly-data-release-january-2021.
4 Sheena Wilson recalled Anne Power as an example of a motivating organiser. A tenants newsletter circa 1976 contained a cartoon (drawn by the Head of Housing at Islington Council, Bill Murray) portraying Anne Power as Joan of Arc mobilising an army of residents (interview Wilson, 2012 and interview Taper, 2013).
5 The total EAP budget rapidly grew from £45 million p.a. in 1986/1987 to £180 million p.a. by 1989/1990 (Hall et al. 2004), so the £50 m spent on DICE between 1989 and 1994 was a significant proportion.
6 The HO published the first British Crime Survey in 1982 as an attempt to uncover the mass of unrecorded crime (of the 11 million crimes estimated to have occurred in 1981, only 3 million were reported) but also the impact of fear of crime. The condition of the physical surroundings was a key contributor to this (Smith 1987) and the British Crime Survey found that vandalism was the most commonly experienced crime and equally the most unreported (Minton 2009).
7 http://www.hta.co.uk/news/posts/hta-at-50 accessed 14 March 2019.
8 The geographical/architectural divide in terminology emerged here. Silver referred to these as colour-coded *maps* rather than *site plans*. McKeown recalled that Coleman 'called everything a map where, for us [Architects] it would be a site plan. Even the drawing of a house was a map' (Interviews Mary McKeown and Peter Silver, 2013).
9 In spite of studying architecture during the late 1980s, Peter Silver was unfamiliar with the controversy around *Utopia on Trial*. His first task on joining the KCL Unit was to read Coleman's and Newman's books (interview Peter Silver, 2013).

10   Paul Wiles, then Professor of Criminology at the University of Sheffield, but later chief scientific adviser at the HO, becoming chief government social scientist in 2007.

11   Michelle Smith was a Housing Assistant on the Ranwell East Estate, London Borough of Tower Hamlets.

12   Both Newman and Coleman cite the size and height of individual blocks as significant variables. But Newman's definition of large estates ranged between 750 to 1,000 homes and his threshold for high-rise was seven or more storeys. While far smaller than any studied by Newman, Ranwell East Estate was the largest of the seven DICE estates and the Rogers Estate the smallest.

13   Sir George Young then Minister of State for Housing and Planning.

14   Although one of Coleman's key abuse indicators was litter the DICE principles did not directly address the problems of refuse stores or chutes in mass housing.

15   For a good review see the SNU's John Farr and Steve Osborn's *High Hopes: Concierge, Controlled Entry and Similar Schemes for High Rise Blocks*, published by the DoE, 1997.

16   This was a contrary position for Coleman to take. Around 1984/1985 she had added a 16th DICE principle, recommending glazed doors over open door apertures. The security technology of the time, entry keypads or key fobs was basic and easy to vandalise. Despite this entry phones (and CCTV) were high on residents' lists of desired improvements.

CASE STUDY
The Mozart Estate: A Laboratory for Defensible Space

Coleman was able to put some of her ideas into practice when she studied the Mozart Estate and her design recommendations were subsequently implemented. (SNU 1993: 45)

The Mozart Estate in the inner London Borough of Westminster exemplifies many of the themes explored in this book: it was the subject of conflicting academic investigation, politically motivated evaluation and waves of practical design changes, illustrating shifting fashions in estate regeneration. It also offers a window through which to view, for example, the media's demonisation of ghettoised estates and the unrealistically rapid policy responses required by governmental election cycles. Significantly, the story of the Mozart Estate emphasises the mobility of ideas on defensible space as a whole, as the various teams remodelling it rejected some approaches to design and crime reduction but accepted others, demonstrating which aspects of the concept emerged as important *through* practice, even if they were not promoted by Coleman herself.

Following McCann's (2011b) identification of policy mobilities as a process of assemblage, the Mozart Estate acted as a landing site, where diverse urban and housing theories for remaking estates were played out and many of the issues discussed were brought together. It is a site, indeed a laboratory, where researchers, architects, crime reduction specialists and policy makers have all tried out their ideas and theories. The 30 years spent rebuilding the estate also spanned the

Defensible Space on the Move: Mobilisation in English Housing Policy and Practice, First Edition. Loretta Lees and Elanor Warwick.
© 2022 Royal Geographical Society (with the Institute of British Geographers). Published 2022 by John Wiley & Sons Ltd.

emergence and mainstreaming of community-led regeneration, reinforcing the importance of involving local residents in the process of altering the estates where they live.

When completed in 1974, the newly built Mozart Estate consisted of 737 houses and flats in 25 medium-rise blocks. Designed by Westminster City Council's Architects Department, it was heavily influenced by Darbourne and Dark's Lillington Gardens in Pimlico, with similar dark red bricks set within concrete frames and tightly packed blocks connected by high-level walkways and 'picturesque' pedestrian routes segregating people from vehicles. Block orientation was dictated by solar paths, departing radially from the adjacent historical street patterns. Phase 1 received a DoE *Good Design in Housing* award and the Mozart Estate was considered to be an exemplar of modern social housing, a more humane alternative to the high-rise tower blocks of the preceding decade.

Yet while the award-winning Lillington Gardens was designated a conservation area in 1990 and continues to be a desirable place to live, the Mozart Estate's history was less positive; from being a design exemplar, it declined in reputation to become a problem estate in less than a decade. Despite the common perception of it as a failed, bad design, the Mozart Estate was where Coleman *piloted* her DICE approaches prior to receiving funding from Thatcher. She saw the estate 'first as a theoretical example and later as a laboratory' (Lowenfeld 2008). It was the DoE's first national Estates Action improvement scheme. It also subsequently became an influential example of community architecture, subject to rolling programmes of rebuilding by the community architects HTA and Abbey Hansom Rowe (now AEDAS) and was cited as an exemplar of community participation in estate management (Queen's Park Forum 2008).

The glowing praise for the architectural design of the Mozart Estate quickly faded, as complaints increased about the construction: within five years flats suffered from damp and poor heating and needed structural repairs (*Paddington Times* 1977). Reports of vandalism, criminal activity and neglectful management increased. Within seven years of being built, the Mozart Estate had become an undesirable and unpopular 'sink estate' (Lowenfeld 2008) (see Figure M.1).

This loss of popularity established a cycle of deepening residualisation, the consequences of Westminster Council's allocations policy and the convergence of housing policy with political forces. The Mozart Estate was initially intended to house the displaced population from slum clearances resulting from the construction of the Westway motorway. The Westway, carving through working-class North Kensington, caused its own social displacement, but housing investment funding delays meant that the motorway was completed well before the Mozart Estate. It then became home for people with few links to the surrounding area, forming a new and isolated community (City of Westminster Council 1993). Council house points-based allocations systems dependent on length of residency and time spent on waiting lists had for years acted as a barrier to immigrants, meaning that in London a disproportionate number of ethnic

Figure M.1   Rapid decline of an award-winning design: Mozart Estate 1982.

minority households were living in overcrowded and poor-quality private rental properties. Local Authority housing officers may not have been intentionally racist but discriminatory practices occurred (what we now call institutionalised racism). If difficult-to-let estates were seen as unacceptable to (usually white) applicants, then they were let to non-whites, or other families whose housing options were fewer and worse. As council housing as a whole became the housing of last resort, unpopular estates like the Mozart were at the bottom of the pile. Additional housing allocation points were awarded to single parents, which hastened the shift of council housing, like that on the Mozart Estate, from 'general needs' to:

> an ambulance service concentrating its efforts on the remaining areas of housing stress and dealing with a variety of 'special needs' such as the poor, the homeless, one-parent families, battered wives and blacks. (Michael Harlow, quoted in Cole and Furbey 1994: 84)

The pressure on affordable council homes in Westminster was escalated by the 1986[1] 'Homes for Votes' scandal' which accused Dame Shirley Porter of gerrymandering, by selling council homes at highly reduced prices in eight Westminster marginal wards with the intention of reducing the number of Labour supporters by 'designated sales' of council homes to potential Tory voters. In a dispute that dragged out over decades, Porter was alleged to have manipulated the Council's policy of 'building stable communities' by keeping council homes empty to improve her party's outcome in local elections. By the 1980s this combination of reducing access to better and more popular council stock, with an unofficial, yet racist, placement policy, resulted in the Mozart Estate's largely low-income Afro-Caribbean demographic makeup.

The Mozart's reputation deteriorated still further during the early 1990s. Sensationalised media reporting of growing crime levels and drug dealing labelled the Mozart 'Crack City' (Spring 1994), intensifying poor perceptions of the estate:

> A young man, pale, scruffy and unshaven, hovers around the edge of one of west London's most run-down estates. He is shifty, apparently suffering from stomach cramps, eyes darting from side to side. The classic crack user out shopping for a rock. (*The Guardian*, 1993, May 22, p. 27)

The vigorous Estate Residents Association complained to the Press Commission that articles such as these stigmatised the families who lived there (cr. Lees 2014 on the stigmatisation of the Aylesbury Estate in Southwark; see also Ferreri 2020 on 'painted bullet holes'), but by 1993 the residents' own rejection of the Estate was clear from the number of void properties and that 63% of residents were on the transfer list to move away (City of Westminster Council 1993: ii).

The Estate's architecture and layout was confusing and oppressive. A maze of overhead walkways led to dark stair towers; ground level garages beneath the walkways were used for criminal activities. Unrestricted access to the blocks and extremely poor levels of natural surveillance resulted in ad-hoc attempts to control access with gates and barriers. The external spaces were harsh and unwelcoming, neglected and litter strewn, with shadowy stairways, hidden corners and confusing dead ends. Architects for the redevelopment recall the Estate's notorious reputation as 'virtually a no-go zone' when they started work there in 1994 (interviews with Peter Oborne and Alan Blyth, 2011). In her 2000 novel *White Teeth* local resident and author Zadie Smith described the turf wars, drugs and stabbings in the alleys and walkways on the Mozart Estate near to where she was brought up.

Initiatives to improve living conditions on the Mozart Estate began in 1983, only a decade on from the estate's completion, when Patricia Kirwan, the Conservative Chair of Housing at Westminster City Council, asked Alice Coleman to apply her design modification ideas to the Estate (Coleman 1985c). A working party selected from Westminster's Housing committee, residents and the police was formed and Coleman was commissioned to provide an expert, outside view, and the first publicly funded application of her design principles (*Architects' Journal* 1985; Anson 1986). The working group was chaired by Estate resident doughty Muriel Agnew, who moved on to the Mozart in 1973, and when the Council ignored her complaints about poor living conditions and anti-social behaviour spoiling the new flats, spent hours knocking on neighbours' doors to form the Tenants Association. She continued to campaign for refurbishment, as well as for facilities such as a community cafe and youth centre. Muriel Agnew was delighted to be coopted onto the regeneration project to work with Coleman.[2] In fact, she was awarded an MBE in 2006, with a plaque to commemorate her contribution to the Mozart Estate unveiled in 2004. She is one of the hidden female transfer agents we pointed to in Chapter 1. Agnew spent three decades campaigning to improve the living conditions on the Mozart, and along with Kirwan (a GLA member and ward councillor from 1977 to 1989) and Coleman whose involvement spanned almost 15 years, formed a triumvirate of women who exerted a long-term influence on the Mozart Estate's form and character.

At the time the working group was formed, Coleman was drafting *Utopia on Trial* (which included a chapter on the Mozart Estate) and used her Design Disadvantagement study to devise a design brief. This brief recommended removing the walkways, reinstating continuous streets with pavements and waist-height walls creating private gardens, providing individual access to most flats and sub-dividing blocks by blocking corridors, separating blocks with boundary fences around each one, and removing green spaces and children's play areas to outside the estate. Coleman's design analysis focused blame on the circulation routes:

> The walkways are the most vicious 'open sesame' making the block vulnerable to outsiders ... All these things add up to the undesirability of anthill designs riddled with walkways, passages with exits, lifts, staircases and ramps. Fortunately, however the worst excesses of all these variables can be cut by a single solution; the removal of overhead walkways. (Coleman 1985c: 137–138)

The high-level walkways provided multiple escape routes for muggers and youths on mopeds, raising fears of other dangerous confrontations. During the riots on the Broadwater Farm estate in 1985, missiles thrown from the high-level walkways ambushed the police and emergency services (Knevitt 1986). Architect Peter Oborne recalled similar attacks on the Mozart: 'there were people being shot, there were fridges being dropped on people from walkways' (interview with Peter Oborne, 2011). Westminster Council followed Coleman's recommendations and in 1986 spent £4 m to demolish four walkways (see Figure M.2). Kirwan and Coleman both appeared on the Thames News[3] promoting the demolition of the walkways. Kirwan also wrote to the then Home Secretary, Douglas Hurd, recommending the removal of walkways from all similar modernist estates (Coleman 1985c).

Coleman was adamant that all the remaining walkways be demolished. However, in 1986, a group of residents asked Bill Hillier and Alan Penn of UCL to

Figure M.2   Demolishing the walkways on the Mozart Estate. (Source: ARH Architects)

evaluate the effects of their removal.[4] Hillier and Penn (1986b) believed removing the walkways would have a negative impact, diverting residents to less overseen and more insecure routes:

> Coleman's design proposals would create a labyrinthine villagey layout with a number of cul-de-sacs which would force pedestrians to use routes adjacent to garages rather than adjacent to the front doors of dwellings. The proposals would divide the estate into isolated, segregated areas which are cut off from the surrounding area. (SNU 1993: 45)

Unlike Hillier's earlier condemnation of Coleman's research methods (see Chapter 5), this critique was based on his essay 'Against Enclosure' (1988) and the detrimental effects of limiting accessibility and restricting pedestrian movement patterns caused by removing the walkways. Overhead walkways were not characteristic of the New York City housing projects studied by Newman and it was Coleman who identified elevated walkways (along with flats raised on piloti or above garages) as one of her *English* 'design suspects' of architectural features that contributed directly to crime. But the evidence for and against walkways was undetermined. Evaluating the demolition of seven walkways from the nearby Lisson Green Estate three years earlier, Poyner (1986) found that their removal did have a positive, but limited, effect on crime, however similar reductions in burglary levels did not materialise. Poyner suggested that alternative alterations, such as the installation of entry phones, might have had a similar effect. When the Council surveyed Mozart households on whether removing or retaining the walkways had been a good idea, they found an even split (42% thinking removal was positive and 47% that it was a bad idea or had made little difference). The main cause of tenants' dissatisfaction, however, had less to do with the alterations than the Council's poor maintenance and cleaning regimes (SNU 1988b).

Despite the debate about the walkways and the unpopularity of some design changes,[5] in 1993 Westminster Council decided to roll-out 'Colemanisation' (see Figure M.3) across the rest of the Estate (*Architects' Journal* 1990). The second phase strategy document demonstrates the extent that Coleman's version of defensible space had been accepted, with the main aim:

> To design out crime and provide a safer, more congenial environment for the residents of Mozart Estate … by creating defensible space for individuals, restoring traditional vehicular and pedestrian circulation to street level, removing the dark, secluded areas which assist criminals; and limiting movement from block to block through the estate. (City of Westminster Council 1993: 7)

Figure M.3   Rolling out 'Colemanisation'. (Source: ARH Architects)

The remaining walkways were to be removed and garages converted to flats with front gardens, bringing activity to formerly isolated routes. Circulation was reconfigured, sub-dividing blocks to achieve Coleman's prescription of a limited number of flats sharing entrances.

Research undertaken by the SNU for Westminster, found that there was 'remarkable disparity' between the problems on the Mozart Estate as perceived by Coleman, based on no empirical research, and those identified by the SNU. They found, amongst other things, 'that more defensible space does not necessarily mean less crime and vandalism' (Chartered Institute of Housing 1989: 13). The findings were so controversial that Westminster refused to present the findings publicly:

> *Housing* can reveal that much of the research contradicts some of the fundamental assumptions on which Alice Coleman's methods rest and questions the wisdom of introducing major changes on the basis of, at best, equivocal evidence. (Chartered Institute of Housing 1989: 11)

SNU said: 'We wouldn't say that Alice Coleman's methods are invalid, just that it's more complicated than she makes out' (SNU's Tim Kendrick, quoted in *Housing* 1989: 13)

Coleman's reappointment after being endorsed by Thatcher, was justified by her claim that the initial alterations 'according to the beat police, resulted in a sudden 55% drop in the burglary rate, which has remained low ever since' (Coleman 1990: 144). This unsubstantiated figure was widely used in the press and architects' publicity (AEDAS 2004; Franks 1990; Hawkes 1991) and by Coleman herself in the proposals for Phase 3 (Coleman et al. 1992). The local Metropolitan Police enthused about the initial changes:

> The first phase of development has from a police perspective made our job easier. It appears to have reduced all aspects of crime and rowdyism and removed the escape routes for rovers. The buildings have taken on a more individual appearance, provided residents with defensible space and encouraged a greater community spirit. (City of Westminster Council 1993: 4)

Yet a second SNU survey in 1993 found that burglary had doubled since 1988, with 74% of Estate residents thinking crime remained an extremely serious problem (SNU 1993). Compared to other Westminster estates the Mozart Estate suffered significant incidence of all types of reported crime with endemic drug dealing and substance abuse (Floyd Slaski Partnership 1993).

Early phases of the Mozart Estate's £27 m regeneration cost[6] was partially funded through £12.6 m of Estates Action grants cross-subsidised from houses built for sale. Funding from the Estates Action programme required completion of *Form B*, a mandatory option appraisal process, which architect Ben Derbyshire scathingly recalled 'purported to be an analytical process' (interview with Ben

Derbyshire, 2012). Similar to Coleman summarising a complex series of design factors into a design disadvantagement score for each block, the whole master-plan was evaluated against 10 variables: the additional homes provided and how it performed against nine, mainly quantitative, criteria. The order of this standard set of criteria, opening with safety and concluding with defensible space as a separate measure, again demonstrates that defensible space was now integrated into mainstream DoE practices.[7] In the *Form B* appraisal, defensible space was defined as providing homes in a defensible environment. This was assessed by counting the number of homes/or routes where the following criteria had been achieved: pedestrian only routes were closed off; ground-floor entrances faced a house or a public through road; homes had a private front garden and/or a private rear garden; garden abutting a garden without rear alleys; blocks were separated from other blocks; reduced dwellings per staircase; parking was overlooked; and public and private spaces were defined. The appraisal counted as positive the number of units experiencing each criterion with no qualitative assessment. Unlike Coleman's design disadvantagement score, no desirable thresholds were set. So, where Coleman's design disadvantagement score recommended a maximum of six dwellings per secure entrance or stair, *Form B* counted as positive any reduction in the number of flats per stair.

Yet it was neither the academic argument about the impact of walkway removal, nor the conflict between fire regulations and achieving Coleman's prescriptive numbers of dwellings per corridor that finally resulted in her removal from the project. As the Price Waterhouse DICE evaluation report was eventually to warn (see Chapter 6), Coleman's proposals were expensive and the increasing costs of the physical changes to estates were becoming unpalatable:

> From 1993 to mid-95 you got a developing sense that applying Coleman's principles in their undiluted entirety to the development was unworkable. (Mozart Housing Manager Dave Bowler quoted in Lowenfeld 2008: 170)

The per-unit refurbishment costs were now comparable to the cost of building anew, with residents experiencing the discomfort of living through disruptive on-site works, without the final outcome of improved internal living accommodation.

In 1997, with Coleman now taking only a minor advisory role, the scheme was again re-evaluated. Alan Blyth of AEDAS assessed the effect of Coleman's alterations, deciding which ones to repeat and how to engage residents more in the design process:

> The early phases brought some lessons learned in practice, so our briefing process involved reconsidering some of the physical measures which had proved to be disproportionately disruptive to residents in occupation or were undiscussed [with residents] or designed by professionals, like the Secured by Design measures. (interview with Alan Blyth, 2011)

Figure M.4   Devising a gradual community-led solution. (Source: ARH Architects)

In an extended resident consultation on a new masterplan (see Figure M.4), AEDAS were commissioned to review the Mozart Estate's layout with HTA designing the homes and spaces. The key concepts behind the revised masterplan were a further attempt to recreate traditional street patterns with more legible routes, the reintroduction of local shops on the estate, as well as application of Lifetime Homes and SBD standards. This altered street layout followed Hillier's broad approach: creating a series of smaller through roads, each tied back into the surrounding street network and providing on-street parking overlooked by the refurbished blocks.

On completion of the final new homes in 2004 the architects reported that tenants showed guarded optimism about the improvements (see Figure M.5):

> I think I'm going to feel safe when the new door entry system is fully functional. (Mozart resident cited in AEDAS 2004)

> People round here are really pleased with the way the place has changed. ... The whole mood of the area has changed – there's definitely less crime and it has quietened down a lot. (Queen's Park resident cited in AEDAS 2004)

Official views promoted a picture of a safe, crime-free area, with the Mozart's rebuilding used as an exemplar case study in Westminster Council's 2009

Figure M.5   Architects combined 'defensible space' and 'secured by design' on the Mozart Estate. (Source: ARH Architects)

Housing Renewal Strategy. However, darker images of the estate continued to be publicised and, in the same year, the Mozart Estate featured in a YouTube tour of 'West London Ghettos'.[8] Then in 2011 following a shooting – the result of inter-gang rivalry, a Mozart Estate resident interviewed by the BBC said: 'They thought by knocking it down and tarting it up, it was going to get better. But it hasn't' (*BBC News*, 30 September 2011). The local press reported a resurgence in raids on crack-houses and cannabis factories on the estate (Dunne and Davenport 2012; Hunter-Tilney 2012), warning of a recurrence of the serious drug-related problems of the 1990s. In 2015 'a 13-strong mob that sold crack cocaine and heroin in a children's playground' on the Mozart Estate were arrested (*Brent and Kilburn Times*, 22 December 2015).

An alternative to the salacious reporting was provided by the CityWest Homes estate newsletter (2012), which covered more prosaic and less headline-grabbing issues, such as parking restrictions, the redecoration of the estate office and community events. The newsletter described the strong partnership between the Residents' Association and the local Safer Neighbourhood police team. Community safety, anti-social behaviour and crime are mentioned, but only in passing amongst news of children's art competitions, advice on housing services and dog training events (CityWest Homes 2012). Nonetheless, a casual visitor to the Estate today sees little indication of these violent incidents. It no longer feels like a threatening, neglected, failing estate; it is tidy and well maintained, with mature landscaping and a mix of residents out enjoying the open spaces.

Most recently the Mozart Estate has featured in London rapper Fredo's drill music as the neighbourhood that shaped his music, or 'the zart' as he calls it (Hunter-Tilney 2019). In 'A Love Letter to the Mozart Estate' Fredo said:

> The media seems to focus a lot on the bad side of Mozart, all the things that come with gang culture – knife crime, drugs, etc. It's not about that. There's so much more to the area. It has a mad sense of community which I think you only fully understand when you live there. (https://crackmagazine.net/article/opinion/a-love-letter-to-the-mozart-estate/)

As these conflicting and fluctuating opinions show, external surface perceptions can be very divergent from the unique positive or negative experiences of the individuals living there. Unpicking these contradictions requires a closer, longer look than either the stereotypes of the inner-city ghetto or successful regeneration scheme would suggest. In Hall et al.'s (1978) study of how the US term 'mugging' was introduced by the British media in the 1970s, manufacturing a moral panic against black youths, they argue that it was the label (not the actual crime itself) that was so invasive as a stereotypical shorthand. There are echoes here with the importing of 'defensible space' from the United States, in that applying a design solution so associated with a particular racialised context (see Chapters 1 and 2) made any perceived threat more visible. On the Mozart Estate, remaking the spaces around homes may have increased territoriality, but it also highlighted the occupancy of these spaces by those who were around the most, had time on their hands or wanted to hang out, and so were most visible. Defensible space further criminalised young black unemployment, conflating policing crime with policing the economic crisis. Poor, black or unemployed social-housing tenants were seen as potential criminals to be controlled, with defensible space and its association with race and crime used as an excuse to legitimise state intervention in their homes.

In conclusion, the Mozart Estate's 30-year regeneration story is complicated: decisions and design changes appeared contradictory, piecemeal alterations were trialled and resulted in unexpected consequences that required remedial work (see Figure M.6). Several times there were calls to rethink the design strategies behind each phase of remodelling. Defensible space theory was used to justify the waves of interventions, either via physically altering the spaces or increasing 'ownership' of the scheme through residents' participation in decision-making. Even when the aims and techniques were the same, critical variances in ideology emerged:

> Hillier and Coleman proposed the same general means – architectural modification – to achieve the same general end – reducing social malaise and crime. The methodological and ideological differences between them – Hillier was an architectural scholar with a modern sensibility, Coleman was a social scientist with a radically conservative bent – are clear. (Lowenfeld 2008)

Where Hillier and Coleman disagreed was on the extent of the physical changes necessary. Was it sufficient to change the external form of the blocks and remove the walkways, or would that just reduce the permeability/accessibility and numbers of people legitimately walking through an estate? Was a better solution to reintegrate an estate into the grain and scale of the surrounding urban fabric? Each of these solutions might be made to work. Practitioners had greater confidence that citing defensible space in their proposals would allow them to make changes than the outcome that would result:

> Physical interventions at Mozart were not too technically difficult – and it worked and still works. Defensible Space was a robust orthodoxy. (interview with Ben Derbyshire, 2012)

That any robustness of the underlying concept is dependent on its flexibility is evidenced in how alternative forms of defensible space emerged independently of Coleman: as placemaking layout in Hillier's traditional street patterns (repeated again in Create Streets' traditional layout, which we discuss elsewhere in the book), as a metric in the DoE's scoring, or the young architects from AEDAS picking and choosing design approaches to apply. All of these were practised on the Mozart. But tellingly while these approaches may have worked well in other locations (see Lea View House estate discussed in Chapter 3), on the Mozart Estate they did not achieve the transformation Coleman claimed. The Mozart Estate may have been a laboratory for design experimentation, but the subjects of these experiments were not lab rats but generations of families whose lives and experiences were shaped by greater structural inequalities. As such, the experimental conditions were too variable, the catalysts for change too many.

Policy mobilities work usually tracks translation between places, yet here we have traced how a policy idea mutated on a single site. So, on the Mozart Estate, the policy mobility of defensible space was not about locational travel, but temporal movement and the flow between repeated or contradictory approaches. The readjustment of the idea by subsequent practitioners (be they geographers, architectural academics or designers) was only possible because the elements of defensible space could be mutated and recombined.

It is possible to see long-term and repeated failure to improve conditions on the Mozart Estate as the outcomes of an unravelling assemblage: Westminster's flawed housing policies, undermined by municipal disinvestment and the dismantling of council housing, and overlaid by heavy-handed implementation of crime and policing policy. Boughton (2018: 217) describes the misunderstood causal chain of blame for the perceived (and actual) decline of mass municipal housing; he argues that council estates should be seen as the *victim* of broader economic dynamics such as increasing worklessness, child poverty, drug taking, the evolution of other crimes and a host of wide societal forces; rather than being the *cause or agent* of their dysfunctionality. Policy mobilities approaches often fail to untangle these hierarchies of policy, practices, time, the direct outcomes from the indirect and how the process of assemblage might unearth hidden disorder as easily as construct a new Utopia. The Mozart Estate case study illustrates that as an estate it has been the victim of wider sweeping forces as much as the scalpel of physical remodelling. Removing the walkways did little to remove the underlying social problems its residents experienced and no doubt continue to experience.

1. Many different versions of defensible space were tried out on the Mozart Estate – defensible space worked in pieces /only for a time, demonstrating policy fragmenting to fit a specific location.
2. Policy mobilities – several rapidly imported ideas failed to provide permanent solutions on the Estate, illustrating failed initial transfer but also that successful onward policy replication cannot be assumed.
3. Policy assimilation – defensible space was part of DoE orthodoxy as demonstrated by its inclusion in *Form B* as the bureaucracy of funding regeneration.
4. Changes to defensible space could not counteract class, unemployment or other structural socio-economic inequalities.
5. Housing policy (in the form of allocations) or active management can positively or negatively override design.
6. The Mozart Estate is a 'nicer place' than it was in the 1970s but as Tunstall (2020) found – most unpopular council estates from the period are today fairly popular mixed-tenure neighbourhoods.

Figure M.6   The laboratory/policy mobility lessons from the Mozart Estate.

Notes

1   See contemporary newspaper reports in *The Times* 14 January 1994, *The Guardian* 15 January 1994, *The Guardian* 20 June 1994 and *The Times* 10 May 1996.
2   'Muriel Agnew obituary', *The Guardian*, 13 May 2010. https://www.theguardian.com/theguardian/2010/may/13/otherlives-obituaries.
3   'Mozart Estate'. https://www.youtube.com/watch?v=K25fNv9VAYY.
4   From the mid-1980s, Hillier and Penn at UCL's Institute of Advanced Architectural Studies used the newly evolving Space Syntax method to analyse the relationship between crime and spatial layout of housing estates for the HO Crime Prevention Unit and SNU (SNU 1993).
5   Coleman's insistence on building over the only football pitch to stop youths congregating was unpopular. Coleman claimed children's playgrounds were 'schools for vandalism' that 'erode controllability by attracting hooligans who vandalise them and make them unsafe for neighbouring tenants'. (Coleman et al. 1992: 2).

6 The £27 m covered all phases of the work until the early 2000s, not just Coleman's changes. The overall figure was £30 m including all costs and fees.

7 The nine *Form B* criteria for assessing a design, in addition to numbers of dwellings, were (in order): improve safety and security; more suitable family accommodation; better located family accommodation; diverse tenure and management; more traditional living environment; reduce number of one bed dwellings; reduce management and maintenance; low numbers of permanent decants; and provide defensible space (Floyd Saski Partnership, 1993).

8 'West London Ghettos'. https://www.youtube.com/watch?v=nD835iqVdTQ.

Chapter Five
Evaluations of Defensible Space

Evaluation, perhaps above all, needs to be *realistic*. The whole point is that it is a form of applied research, *not* performed for the benefit of science as such, but pursued to inform the thinking of policy makers, practitioners, program participants and public. (Pawson and Tilley, 1997: 3)

Claims have been made about the success of various crime prevention initiatives. The evidence put forward to substantiate these claims varies from, at one end of the spectrum, the subjective judgement of those closely involved, to, at the other, seemingly comprehensive and exhaustive evaluations carried out by independent researchers. (SNU 1993: 156)

Just as the previous chapters show how politics was central to the funding and implementation of defensible space ideas, evaluation of their success was an equally politicised activity. Theodore (2019) talks about 'evaluation science' as part of policy mobilities interest in 'techniques for learning', but it is more than this. The evaluation of a policy examines its aims and objectives, implementation and impact in order to underline (or not as the case may be) its merit and worth. The success or failure of a policy, of course, is both subjective and slippery. One consequence of the supremacy of political authority was the distortion of the evaluative process, as Flyvbjerg (1998: 15) wryly notes: 'The result seems predetermined and the evaluations … become more ritual than real'. This chapter considers the evaluation of DICE in the wider context of DoE evaluation and the SNU's meta-analysis of crime prevention on council estates. The DoE's perspective on policy and performance evaluation in the early 1990s stemmed from

Defensible Space on the Move: Mobilisation in English Housing Policy and Practice, First Edition. Loretta Lees and Elanor Warwick.
© 2022 Royal Geographical Society (with the Institute of British Geographers). Published 2022 by John Wiley & Sons Ltd.

Judith Littlewood's 1992 report *Policy Evaluation: The Role of Social Research* (Doig et al. 1992). As Head of the DoE's Research Division, Littlewood encouraged widespread evaluation of policy as part of the public sector quest for effectiveness, efficiency and economy.

In 1993 the DoE published a significant and comprehensive evidence review, *Crime Prevention on Council Estates* (see Figure 5.1). Commissioned by the SNU under the supervision of DoE research manager David Riley, it gathered examples of crime-reduction interventions on housing estates. These were categorised as: design-led, management-led, security-led, social development-led or policing-led interventions. A section titled *Urban Design and Deviance* described Oscar Newman's pioneering New York research, Alice Coleman's *Utopia on Trial* and the redevelopment of the Mozart Estate in London. DICE is included as a design-led example, with various case studies of PEP- and EAP-funded schemes included as best practice of management-led local security and social improvement projects. This was despite programmes such as PEP, which had broad aims, being very difficult to evaluate in terms of the specific impacts on crime. The evaluation of PEP in Tower Hamlets, for example, was:

> confounded by the simultaneous workings of changes in local government housing policy, the initiation of 'right to buy' legislation (permitting tenants to purchase their homes), attempts to initiate a HAT (housing action trust) and the pervasive impact of social change wrought by the Docklands scheme. (Downes and Rock 2007: 305)

The tone of the 1993 DoE report was pessimistic and cautious, recognising that despite extensive practical crime-prevention activity throughout the 1980s, crime rates on council estates had continued to rise into the 1990s.[1] The report contained reservations over the quality of information available, the resultant inconclusive evidence which 'does not allow for judgments one way or the other' (SNU 1993: 3) and long sections on the limitations of the report. A (self-evident) conclusion was that assessing the effectiveness of a crime-prevention scheme hinges on the validity of observers' interpretations (SNU 1993: 3), yet these observers rarely include the full range of community stakeholders. The report reinforced the importance of the experience of living in an area, arguing that residents must be the ultimate judges of the success, or not, of crime reduction schemes. Crime prevention is only one aspect of a resident's quality of life and one of many aims of estate-regeneration schemes. Nonetheless, it can dominate thinking, particularly amongst (design or evaluation) professionals and it is misleading not to incorporate judgments of the community on this issue.

In addition, the SNU report proposed a framework for evaluating crime-prevention initiatives (SNU 1993: 31–32), advising on the forms of evidence required for a fair and robust evaluation of a policy intervention. Conceding that evaluations can be expensive, the SNU argued that when costs constrain undertaking a comprehensive evaluation, applying a consistent rigorous basic framework becomes *more* essential

Figure 5.1   The most successful crime-prevention projects have involved some or all of these key measures as part of coordinated packages. (Source: Review of SNU *Crime Prevention on Council Estates* report, in Estates Action Update newsletter, summer, 1993: 10)

to highlight the limitations of an evaluation's findings. The framework suggested that four forms of evidence are required: i) evidence that crime had actually been reduced, in addition to residents thinking the problem had reduced; ii) evidence that the initiative was responsible for any alterations; iii) evidence of which individual measures accounted for any changes; and iv) evidence of the permanence and replicability of the effects. The same DoE civil servant, David Riley, managed both the SNU review and the DICE evaluation, which we discuss later, making it reasonable to read these expectations across both projects.

The SNU meta-analysis highlighted difficulties identifying which individual factors led to success. They found successful schemes shared common approaches with less successful ones. Referring to the opportunistic model of crime which focuses on individuals rather than societal causes (see Herbert 1982; Kitchen and Schneider 2005; and Chapter 2) the report notes that initiatives may have applied certain approaches not because of their proven efficacy but merely because they were fashionable:

> This reflects the existence of a kind of orthodoxy of approach, a general acceptance of a particular model of crime prevention, which emerged in the late 1970s and 1980s and which meant that many initiatives have adhered to or at least paid lip service to certain ideas or principles such as multi-agency working or resident consultation. (SNU 1993: 3)

The SNU were disappointed to identify only eight design-led evaluations (including their own appraisal of the Mozart Estate remodelling – see the case study following Chapter 4). Given this lack of evaluative evidence, the number of expensive design-led remodelling projects seemed surprising. SNU anticipated that the DICE evaluation, which was announced in 1991, would properly clarify the value of design interventions:

> It is hoped that the Department of the Environment's *Design Improvement Controlled Experiment* currently underway settles at least some of the arguments about the impact and cost effectiveness of design measures. (SNU 1993: 103)

They also refer to the DICE evaluation in the concluding paragraph, as part of a plea for evaluation becoming a condition of scheme funding and greater collaboration between practitioners and researchers. This added to the weight of expectation for DICE to be thoroughly and honestly evaluated (SNU 1993: 165).

The DICE evaluation began with a highly charged politicised selection of consultants, Price Waterhouse, who oversaw the process, and we show how they set out with the DoE to suppress Coleman's ideas. We discuss the tortuous relationship between the evaluation team, the DICE unit and the DoE who could not be seen to be countermanding the Prime Minister Margaret Thatcher, until she left office. We then move on to Coleman's own evaluation of DICE, and compare the two. These two exercises were at divergent ends of the evaluation spectrum. One evaluation was well funded, gathering extensive (almost excessive) quantities of data, the other a shoestring exercise constrained by resources and methodological intent. Both were ambitious in their aims, yet neither convincingly achieved the reach or impact that their respective authors hoped for.

The notion of evaluating policy emerged in its modern form in the 1960s and 1970s and was associated with the post-war welfare state and the expansion of government public programmes. It grew out of the appraisal of welfare programs in the United States but soon spread to other countries, including the United Kingdom. The history of policy evaluation in government practice shows that it has followed different pathways in different countries. Evaluation itself became the 'mantra of modernity' and underwent a 'quest for scientific status' (see Pawson and Tilly 1997, for an excellent history of the evaluation movement), justified by a supposedly rational evaluation of programmes and projects. Evaluation was intended to be a process of objective, systematic and empirical examination of the effects of policies and/or public programmes or projects in terms of the goals they sought to achieve. It was linked to the idea of society advancing through research-driven policy-making (evidence-based policy making as we know it now); those policies that failed evaluation were thus poor and to be dropped, those that passed evaluation deserved to be continued. From the 1980s onwards in the United Kingdom and the United States policy evaluation was increasingly used by those critical of welfare-state government programmes to suit their own political goals and indeed to argue for the elimination of certain government pro-grammes. Thatcherism and Reaganism brought with them a new, neoliberal era of policy evaluation dominated by a different philosophy of governance:

> Government had taken on more than it could handle .... decentralized decisions of the market-place should whenever possible replace the inevitably inadequate plans of central or local government. (Pollitt 1993: 12)

The neoliberal era of policy evaluation was what Fischer (1995) calls a 'tech-nocratic world view' that claimed the value neutrality of evaluation and the authority of the policy expert. It was positivistic and economistic. Theoretically it should be inevitable that a positive evaluation outcome would/could improve a policy's mobility and take-up. But recent work on the mobility of policy failure, and policy worst practices (see Lovell 2019), intimates that this is not always the case. The evaluations of defensible space ideas and DICE in particular were important in validating or disproving the 'science' of defensible space, but ulti-mately, and perhaps unusually, they did not affect the resilience and uptake of defensible space ideas, which we discuss in Chapter 6. Nonetheless, in social sci-ence there remains an assumption that a good evaluation of research validates it as good, high-quality science that deserves replication, but the story here is not as straightforward as this diagram would imply (see Figure 5.2).

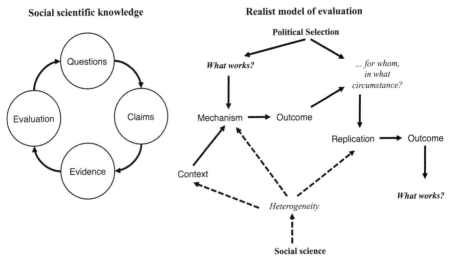

Figure 5.2   Comparing the simplistic circuit of social scientific knowledge to the complex realist model of evaluation. (Source: After Sherratt et al. 2000: 18; Pawson and Tilley 1997)

Evaluations of Defensible Space in PEP and EAP

Before turning to the evaluation of DICE it is worth looking briefly at evaluations of the PEP and the EAP. As introduced in Chapter 4, the PEP was launched in 1979 and grew out of research carried out by the DoE into difficult-to-let housing. The research findings that factors contributing to hard-to-let properties were not the same on every estate, led to experiments in local, intensive management at 20 estates across England. These PEP experiments were intended as the first of a series of initiatives designed to develop good practice in relation to estate management and regeneration. PEP was rapidly superseded by the EAP in 1985 when the then Minister of State for Housing, Sir George Young, established the UHRU to tackle the 'crisis of confidence' in many Local Authority managed estates (Mawson et al. 1995). Like PEP, the EAP was funded by 'top-slicing' a proportion of the existing HIP and inviting submissions from Local Authorities for supplementary capital borrowing approvals. Initially, eligibility was restricted to 69 Local Authorities (36 Metropolitan Districts plus all London Boroughs), but in 1987/1988 the EAP was extended to all housing authorities (DoE 1996). The initial annual investment in EAP (1986/1987) was £45 million. This rapidly rose to £180 million by 1989/1990 (Mawson et al. 1995) demonstrating the commitment to rolling out the approach. The types of work permissible for EAP included the provision of defensible space, external improvements to dwellings, security measures (e.g. concierge schemes) and environmental improvements including estate layout (DoE 1996).

As a programme PEP was subject to multiple evaluations, most concluding it was modestly effective, but that better housing management was not sufficient in itself to ensure sustained improvement (Power 1984a, 1987; Glennerston and Turner 1993). This view was reinforced when conditions on estates were re-visited over subsequent decades (Power and Tunstall 1995; Tunstall 2020). In comparison, formal evaluation of EAP was initially sparser (Pinto 1991), with a greater focus on the formulation of the programme. The PEP evaluation most relevant here is *Housing, Community and Crime: The Impact of the Priority Estates Project* (Foster and Hope 1993). Figure 4.1 shows the timing of this PEP evaluation overlapping with both EAP and DICE initiatives. For this, sociologist Janet Foster used ethnographic interviews, observation and focus groups on two of the PEP estates – she spent 18 months in one estate in Tower Hamlets, London and 12 months in one in Hull. A questionnaire was undertaken on the experimental and control estates, before and after the PEP changes. There was a growing sense at this time that a good understanding of the social processes underlying the formulation and implementation of crime-prevention efforts was important. This underpinned the perception that qualitative techniques were more useful than quantitative techniques in evaluating impacts (Crawford and Jones 1996). Foster used both methods to conclude that there had been a reduction of crime on the estates where PEP approaches had been applied (Foster and Hope 1993; Hope and Foster 1992). PEP was founded on two related initiatives: first, decentralisation of housing services including repairs, caretaking, rent collection or lettings to a very local level, plus a coordinated effort to improve the physical conditions on the estates; and second, involvement of tenants in the management of their estates with the aim that social or criminal problems, once controlled, would not re-emerge. Foster was not, however, interested in all aspects of the PEP and focused mainly on its impact on crime:

> in order to unravel the inter-relationship between the various possible causal factors, explanations may also be needed of the internal dynamics of community change on estates – in other words a chart of the 'crime career' of particular estates over time. (Foster and Hope 1993: 10)

At the request of tenants, PEP schemes often upgraded the security of homes, applying defensible-space-type changes to the immediate surroundings of the dwelling. Yet these physical improvements were found to enable occupants to exert control only erratically. For example, entry-phone systems needed resident cooperation to operate effectively, for they did little to counter fear of crime if residents were afraid of neighbours within their blocks.

Foster's team were not practically involved in the delivery of the programme and were learning about PEP through their evaluation. Their positive evaluation identified four ways that PEP measures potentially inhibited crime-generating processes (Foster and Hope 1993: vi; Pawson and Tilly 1997):

1. Levels of security and degrees of defensibility of space create varying but modifiable degrees of opportunity for crime. Therefore, crime may be reduced by 'creating better dwelling security' and more 'defensible space' through PEP physical improvements.
2. Physical neglect provides significant cues to criminals that the neighbourhood is not cared for and hence no one is likely to intervene and stop them offending. Thus, crime may be reduced by 'halting the spiral of deterioration, tackling vandalism, caretaking, cleaning up the estate, thereby reducing "signs of disorder" and fear of crime, and signifying the estate is well cared for'.
3. An estate in which residents do not have a stake lacks people who are interested in creating and maintaining its security, as those who can do so leave. Crime may be reduced by 'investing in the estate so that residents will develop a positive view and thus a greater stake in their community, and a greater expectation of law-abiding behaviour'.
4. Lack of effective formal and informal social control is conducive to crime and frequently weak in run-down housing estates. Hence crime might be reduced by 'increasing informal community control over crime both through surveillance and supervision by residents'.

These approaches elaborated on Hope and Dowds' (1987) proposed model for how residents' informal control over their estates might reduce crime (see Figure 4.5). Changes in estate layout and responsibility for management that increased interaction and involvement with neighbours via leadership, friendship networks or other methods would reduce mistrust of neighbours, reinforcing residents' positive sense of their estates. Establishing norms of behaviour and collective expectations for when, and how, to intervene would both positively impact on fear of crime.

Foster and Hope's (1993) ethnographic methodology facilitated a detailed knowledge and understanding of the two estates and their social networks, and the social community component of crime reduction. Yet, the non-randomised matching of 'experimental' and 'control' estates could only be considered a quasi-experimental evaluation, particularly as by the end of their study period only about a third of the estates had undergone defensible space improvements. As the authors admitted, the experimental sample of two PEP estates made their findings difficult to generalise to other estates (Clarke and Dawson 1999: 123). Ultimately, their evaluation found that successes were only partial: either limited to localised areas on the estates, or particular resident groups. The PEP model of multi-agency and community collaboration depended as much on the quality of the professionals involved as the initiatives or management structures proposed. A further obstacle to the approach was community instability arising from resident turnover, residualisation and the desire to move away from 'stressed' estates. Stability is required to enable community motivation to organise social control against criminal activity on an estate (Power 1987). An essential factor in embedding social control is improving residents' perceptions of their estate often through an increase in activities outside their home and interacting with neighbours (stimulating both informal, natural surveillance or participation in community management). These indirect and complex interactions between the factors made evaluation and attribution challenging, a common problem for all of the regeneration programmes:

> The PEP approach is not, to be frank, much better informed about causation than *Utopia on Trial* … [It is] eclectic, and mainly symptom centred. (Spicker 1987: 291)

PEP was highly popular within the DoE but its limitations were recognised. Such a comprehensive, estate-based approach required intensive input and outside support from DoE consultants who were only available for a limited period of time. This made it hard to disentangle the local community organisation from more centralised council activities and, as Foster and Hope's (1993) evaluation pointed out, highly susceptible to local political interference. The lack of legal or financial frameworks made its long-term local finance model insecure, requiring Local Authority reorganisation and restructuring. When EAP was launched in 1985, it tried to address some of these issues. Funding was available for a mixed range of activities to encourage authorities to improve their management structures, tenant consultation procedures and to help improve their estate security and environment. Yet these measures to counteract the housing authority's management difficulties were set alongside measures that promoted the contraction of the council housing sector, such as funding to help the creation of housing cooperatives, transfer to Housing Trusts and private sector involvement (Pinto 1991: 106).

While EAP was funding security and other environmental improvement works or tenant-led estate-based management to improve residents' living conditions, equally it was also aiming for greater diversification of tenure and homeownership on estates through the 'right-to-buy' scheme and private sector involvement in maintenance and construction projects:

> The key ingredients [in Estates Action] are responsive, comprehensive and *effective local management* ... including the development of tenant management cooperatives to encourage the *fullest possible tenant involvement* ... We encourage local authorities to *involve the private sector* in the upgrading process, both to generate additional resources and to widen the tenure mix. (DoE 1986: 6, emphasis added)

There was also a far greater emphasis in EAP, than in PEP, on other social initiatives such as employment or homelessness prevention programmes. Despite the similarities between elements that PEP and EAP would fund (and there was overlap where EAP funding paid to support PEP activities) this significant additional policy emphasis on innovation and private-sector engagement was critical. Whenever opportunities emerged, EAP would encourage the disposal of empty homes for sale, or where there were tenants, the DoE Unit would explore the scope of disposal to trusts or cooperatives (DoE 1986). Despite the programme's initial emphasis on these forms of transfer to Tenant Managed Organisations or housing associations, in practice very few schemes did so. It took the 1988 Housing Act and the subsequent LSVT programme to kick-start the wholesale transfer of former council homes to housing associations.

Pinto's evaluation of the first three years of EAP (423 schemes from 1986 to 1989) evaluated the extent that building 'facelifts' were the most popular aspect of renovating unpopular estates. In these early years a third of all funded schemes were either for landscaping, clearing rubbish and tidying up grounds (176 projects), or improving the exterior appearance of dwellings (225 projects). These aesthetic improvements took a substantial proportion of EAP resources, fuelling the suspicion that the programme was priming estates for sales and privatisation. Capital investment in security and environmental works was also a central tenet of EAP and a large proportion of the budget was spent on security works such as entry-phone schemes; improved security measures such as doors, fencing and better lighting; and setting-up concierges (30%, 19% and 23% respectively of the EAP budget between 1986 to 1989) (Pinto 1991: 170).

EAP placed emphasis on 'innovation' and trying new approaches. Yet at the time of Pinto's evaluation he considered that most EAP schemes only really funded the 'defensible space' elements, and the concierge schemes could be considered as being relatively untried. He notes that Alice Coleman's work and DICE generated much discussion amongst Local Authorities applying for EAP finding and that as such, defensible space was given a higher profile in the EAP 'themes' (Pinto 1991: 186). There was though a growing interest from Local Authorities in using EAP to fund concierge schemes – this was particularly popular for tower blocks – and an increasing number of authorities began to experiment with this approach. In 1986/87 seven concierge schemes were initiated, increasing rapidly to 20 the following year, around 11% of all the schemes (Pinto 1991: 170). The other security initiatives undertaken were, however, more commonplace ones, which Local Authorities maintained they would have undertaken as part of planned maintenance if they had the resources to do so. Pinto's review questions EAP's efficacy, wondering whether the – by now – well-established PEP tested and proven solutions might be more relevant and effective (Pinto 1991: 284).

As with DICE, there was a tension in EAP identifying which estates to target, with a frequent criticism that unpopular estates were being tidied up in preparation for the disposal of council housing stock. The programme struggled with weak objective measures to identify which estates would benefit most, rather than schemes that would deliver the greatest number of policy outcomes. Local Authorities were expected to contribute half of the capital costs leading to delivery bias against those authorities with fewer resources but relatively greater need. The *Estates Action Handbook* (DoE 1991), with its principal concern for public cost benefit and emphasis on value for money, highlights this inequitable balance of benefits to those producing or delivering the schemes (additional rents collected by the landlord) against those experienced by occupants. Monetary benefits for crime reduction may accrue to residents, for example, reduced insurance costs as well as less damage or losses from crime itself (DoE 1991: 148). Nonetheless, non-monetary benefits are often perceived to be more significant: the reduction in fear of crime and anxiety, increased participation in recreation and social events, increased business activity, reduced stigma or employer confidence in residents as potential employees etc. (SNU 1993: 150).

EAP's quantitative outcome-driven methodology was a consequence of nationally conceived solutions to nationally perceived problems. The SNU pointed out that the nature of crime problems may worsen even if the number of incidents reduces, and identified the importance of measuring both frequency and severity of local crimes. The *Estates Action Handbook* was also particularly weak on how to monitor or aggregate localised external benefits. Unlike DICE, no mention is made of gains to other departments, agencies or wider local stakeholders such as businesses. Similarly, benefit cost savings attributed to staff time from reduced repairs may equally have been achieved though concurrent productivity measures or effective management practices.

As a programme EAP was criticised for being too inflexible in terms of the projects and the works to be supported. Overall, the success of EAP was dependent on there being alignment to the perceived needs experienced by a particular Local Authority and work they would have wanted to undertake anyway. This finding adds complexity to the work of Marsh and McConnell (2010: 571), who distinguish between different types of policy success and failure: process, programmatic and political.

Price Waterhouse Evaluate DICE

> The evaluation looked at both financial and qualitative issues and assessed the long term durability of the changes. The results were mixed, showing no clear pattern of success or failure. Overall the consultants declared DICE projects to be no more or less effective than contemporary Estates Action schemes. (Towers 2000: 116)

An evaluation of DICE was necessary because the DoE needed to demonstrate the impact of its £50 million investment, and Alice Coleman's receipt of funding was conditional on her participation in an independent, in-depth assessment of the impact, costs and benefits of the initiative. The evaluation became a complicated, costly exercise combining data gathered from multiple sources, such as the police, social services, as well as large-scale longitudinal surveys of DICE-ed estate residents. David Riley, the government client, was a social research specialist who, like Sheena Wilson, had worked at the HO before moving to the DoE. At the time of DICE, he was a relatively junior civil servant (later to become Chief Social Researcher, Head of the Analytical Services Division at the Health and Safety Executive) progressing rapidly by managing significant joint department evaluations including the SNU report (1993) and PEP evaluation (Foster and Hope 1993). Evaluation was a typical expectation for most publicly funded programmes (Wells 2007) and was also a condition for Local Authorities receiving DICE funding. However, David Riley felt that Alice Coleman agreed to the evaluation 'very dismissively without any intention of paying attention to it' (interview with David Riley, 2011), referring to early correspondence between Alice Coleman and Sir George Young.

There was a feeling amongst the civil servants that Coleman's evidence for such design-led interventions was insubstantial, which led to the Head of Division for Research, Judith Littlewood, advising the then DoE Permanent Under Secretary Terry Heiser that the proposed research was inappropriate. Riley summarised his briefing to Littlewood as 'basically say, you do it over my dead body but if you do it then it's got to be properly evaluated' (interview with David Riley, 2011). He justified the extensive evaluation on the basis of exposing this implausible evidence base:

> It was nonsense to start with. You knew perfectly well what the evidence would be. Our concern was to cover all the bases so we couldn't be accused of overlooking something. It seemed such an improbable mechanism to produce any benefit,

especially since you managed to piss the tenants off enormously in the process. (interview with David Riley, 2011)

Despite describing DICE as a full-scale pilot project, the DoE were anticipating that the positive outcomes from these kinds of design interventions would not be commensurate with the high level of investment, and as such did not intend it as a trial run for extensive roll out. One of the Price Waterhouse evaluators, Katrine Sporle, reiterated the worry over the expense of similar DICE-like schemes proliferating. One intention was 'to ensure that Coleman could not run away with public money and start doing this all over the place' (interview with Katrine Sporle, 2011). These pre-evaluative, political (with a small p), intentions were, of course, not commensurable with the positivistic notion of objective evaluation.

Coleman's earlier derogatory comments about the DoE in *Utopia on Trial* and the poor fit of her proposals alongside other more favoured regeneration approaches such as local management offices, concierges or resident caretakers (DoE 1981[2]), added to the suspicion of DICE within the Department. This negative DoE view of Coleman's work may not have been communicated formally to the evaluation team, yet Sporle recalls that they were well aware of the departmental political tensions surrounding it, alongside the Prime Minister's support. They were also aware via John Harvey that the number of Local Authorities enthusiastically applying for DICE funding was causing hostility within the Estates Action team (interview with Katrine Sporle, 2011). Here the different operationalisations of defensible space within different housing programmes came into conflict over funding and this played out informally in the evaluation.

In June 1989, having commissioned Coleman's research, John Harvey moved from the DoE to Price Waterhouse Management Consultants. Acting as a 'transfer agent' between the DoE and Price Waterhouse, he passed on his knowledge of the DICE project's background and the policy context into this commercial setting. Once there, Harvey helped the young project manager, Katrine Sporle, draft the tender proposal for the DICE evaluation. Price Waterhouse competed for the tender, were appointed and brought together a sizeable team of evaluators including the market research firm MORI; all project managed by Sporle.[3]

Forming a steering group with Coleman's participation was a sensitive first step in shaping the evaluation. David Riley remembers cannily selecting the steering group members:

I was mindful of Alice Coleman's rather slippery use of stats. So I invited an eminent statistician. Not letting her wriggle off the hook as it were. It had to be open and transparent and done in her presence. Having this steering group with independent experts on it turned out to be the right way forward. She did cavil at a number of the findings, but they were validated by independent experts. (interview with David Riley, 2011)

These experts were: statistician Sir David Cox from Nuffield College, Oxford; Tim Hope, the criminologist and 'astute methodologist' from Keele University;[4] Chief Inspector Brian Hewitt of the Stafford Crime Prevention Centre; and Paul Wiles, then Professor of Criminology at Sheffield University. Representatives from the DoE, Department of Health and the HO also contributed. Professor Wiles, who later spent nine years as the Chief Scientific Adviser at the HO, becoming Chief Government Social Scientist in 2007, was a particularly combative and outspoken critic of the DICE programme and Coleman's research from the start. Nonetheless, the evaluators appreciated his expertise, finding him 'a marvellously knowledgeable expert, able and sensible' (interview with Katrine Sporle, 2011).

Coleman was, perhaps understandably, displeased with the set up for the evaluation. An article in *Building Design* sensationally described her as 'seething' at the imposition of Price Waterhouse as external monitors for the programme (Baillieu 1991b). Coleman was, perhaps understandably, suspicious of the report's impartiality as its authorship was attributed to John Harvey. Coleman fumed 'We will do it, it will be a great success; they [Price Waterhouse] will write it up and say it is a great failure' (Baillieu 1991b: 1). Harvey later relocated within Price Waterhouse to minimise conflict of interest and accusations of bias arising from his involvement evaluating a project that he had commissioned.

Price Waterhouse's evaluation was a flawed evaluation from the outset. In the early 1990s there were weaker civil service mechanisms for monitoring the progress of research studies and ways to halt irrelevant projects. The first Treasury guidance on policy evaluation was published in 1988 (HM Treasury 1988) and in 1992 Judith Littlewood co-authored the DoE guide *Policy Evaluation: The Role of Social Research* (Doig et al. 1992), but formal guidance on integrating evaluation into government level decision-making was relatively slow to evolve, although it had been an implicit element of policy-making for several decades (Strategic Policy Making Team 1999). *The Magenta Book* (Government Social Research Unit 2003[5]) eventually provided a practical guide to action research and evaluation techniques. In parallel the first Treasury framework for appraising all kinds of investment *The Green Booklet* was published in 1984 (HM Treasury 1984). This was eventually replaced by the Treasury *Green Book* (2003), which included non-monetary assessments for design (in the form of the Design Quality Indicators), amongst more usual economic evaluation.

Even though formal policy evaluation was in its infancy, practical project decision-making guidance was available in the form of the DoE's own *Handbook of Estate Improvement Vol. 1 – Appraising the Options* (1989), which applied the Treasury's economic techniques to housing estates. This and subsequent editions of the *Estates Action Handbook* provided a guide for option appraisal for all Estates Action schemes. Although it concentrated on cost benefit and economic analysis, the handbook did attempt to record the value of more intangible benefits, such as quality of life, fear of crime and aesthetic preferences. But the question was asked:

> is a reduction in fear of crime, for example, worth a sacrifice of net quantifiable benefits? If so, what are the limits to this sacrifice? (SNU 1993: 119)

So, as well as the DoE having fewer decision-making mechanisms that they might have used to halt the study, there were also powerful drivers for the consultants to continue with the evaluation, regardless of their opinions:

> We went on for months and months because we had to. It was one of those situations where Price Waterhouse weren't going to blow the whistle on it, they were making a fortune out of it. Later on there were some mechanisms that might have got you out of that. In other words, if one year into DICE somebody was in the position to go to the Minister and say 'this is barking frankly, we're throwing money at this and it's not going to work', then the whole thing might have been derailed. David Riley, who was a fairly junior research officer at that stage, was managing it. Judith Littlewood was quite influential but there were many other things that she was trying to do. In any case, the problem with Judith is that everybody knew that she was a great supporter of an alternative answer to this problem [PEP] and that kind of discounted the advice she was giving, if she wasn't careful. The fact was, everybody was bloody terrified of the Prime Minister. (interview with Paul Wiles, 2011)

Moving forwards, fixing on a suitable evaluation method was challenging. The basic intention was straightforward – collect information prior to the start of refurbishment works on the site and return a year after completion. Similarly, analysis was planned at two scales: 1) assessing the impact on an individual estate relative to its own neighbourhood (and a local control estate); and 2) a comparison of the overall pattern of outcomes against a pool of all the control estates. However, the comparison control estates were chosen first for their similar block form (regardless of estate size) plus a very rough match of socio-economic classification of the residents. This simplistic evaluation design controlled poorly for bias (particularly arbitrary geographical variables, such as the Local Authority). The title 'Design Improvement *Controlled* Experiment' (DICE) illustrates Coleman's aspiration for a pseudo-scientific comparative approach, dictating the use of estate controls and controlling for as many variables as possible (interview with Alice Coleman, 2013). However, as Chapter 4 has shown, potential estates were selected based on a preconceived idea of whether they could demonstrate the success of her DICE principles. Moreover, by the 1990s this use of experimental-control design was generally considered unsuitable for the investigation of social groups in a community setting, meaning the Price Waterhouse approach was little more than a structured case study comparison:

> So purity of experimental design was almost impossible, given Alice was in charge of the actual implementation and choosing the areas she was going to implement it in. You had the whole thing contaminated right from the beginning. (interview with Paul Wiles, 2011)

In fact, Riley was very concerned about the inherent weaknesses of the experimental design and the increasing complexity of the analysis attempting to control for the many variables that emerged in response (interview with David Riley, 2011). For both evaluation design and analysis, the aim was/is simplicity:

> A good experimental design is a simple one, not complex. It's the same as statistical analysis. If you have to do all sorts of fancy statistics, you've got bad data. If you've got really good data, good strong experimental design, the stats are easy. You only get into fancy modelling and stats when your basic design is not very good. (interview with Paul Wiles, 2011)

Riley described Coleman's measurement parameters (graffiti, cigarette butts and litter) as 'rather trivial'. While these might act as diagnostic indicators, he felt trying to measure the impact of building-design changes with such inconsequential outcome measures was ineffectual, particularly when the project was interested in crime, fear of crime and social behaviour. Coleman's measures could not explain the benefits to residents, so the evaluation needed to gather a greater range of behavioural and attitudinal data (interview with David Riley, 2011). As such, the ambitious quasi-experimental evaluation design devised by Price Waterhouse resulted in an increasingly convoluted process of data collection and analysis.

One weakness of defensible space as a concept, and Coleman's application of it in practice, was the breadth of influence/impact she claimed for it. Attempting to assess these diverse influences to evaluation swelled; there was too much data and the findings were too broad. The Price Waterhouse analysis tried to identify effects against eight broadly defined areas of impact and gathered data for all these areas:

- crime and incivilities (which included locational analysis of crime incidents, as well as fear of crime and incivilities);
- housing management (assessing resident satisfaction with management and repairs);
- social fabric and community (which included indicators for social control, such as the presence of strangers on the estate);
- upbringing and control of children (based on parents' awareness of their children's activities and locations for play),
- socio-economic conditions;
- health of residents;
- environmental conditions (this most closely echoes Coleman's measures but assessed the *extent* (not only presence) of graffiti, litter and damage and also residents' general satisfaction with the environment); and

• desirability of the estates (which included satisfaction with the dwelling as well as its surrounding area).

Applying more sophisticated mapping than either the Design Disadvantagement study, or the DICE reports, the evaluation recorded the location of anti-social behaviour and criminal incidents, as well as Coleman's graffiti or littering variables. However, in a pre-geographic information system, pre-computer era, the techniques used were basic, with data collated as coloured stars stuck onto large-scale drawings, sellotaped onto A1 sheets of paper:

> We had to send mappers out. Every part of the estate had a number attached to it. When we asked an interviewee, 'What happened to you on the estate?' They had to say precisely where on the estate it happened to them and the interviewers had to record on the questionnaire these numbers, so that everything had to be taken down. So not only were the dog turds themselves subject to quality control – size and density – but everything that happened anywhere was assigned to these location numbers. What would we do with all this data? It was madness! (interview with Toby Taper, 2009)

None of this spatial investigation was used in the final report and it appeared to play little part in the analysis. There was consensus amongst the evaluation team that too much data and information was collected, with much of it unused (interviews with Katrine Sporle, 2011, Toby Taper, 2009, and David Riley, 2011):

> I suppose the number of issues covered in the evaluation were probably more than strictly necessary. But it was belt and braces. There'd got to be no stone left unexamined. I think genuinely there was interest in unintended benefits. Because a lot of money had been spent on these estates. They did look different afterwards. There may have been issues that weren't predicted by Coleman that were genuine outcomes of the intervention. (interview with David Riley, 2011)

The evaluation's five estates, each with three controls, were assessed pre- and post- refurbishment between 1991 and 1995.[6] The following two years were spent analysing the considerable quantity of information gathered, resulting in a short, final report and even shorter research summary, published in 1997.

The cautiously worded findings covered two themes: the impact of DICE on individual estates and a value-for-money assessment. The report supplied a detailed evaluation of the five estates with calculations of a wide range of variables reporting on both monetary and non-monetary outcomes and the long-term sustainability of the effects. The evaluation report listed the objectives of the DICE programme:

> In particular the evaluation assessed the extent to which the wide range of objectives predicted by Professor Coleman to flow from the remodelling of poorly-designed design estates were actually achieved. These were expected to include:

- substantial reductions in maintenance costs,
- easier and more efficient estate management as a result of fewer complaints,
- easier and cheaper estate management through a lower tenant turnover rate,
- reductions in rent loss as a result of shorter vacancy periods and more satisfied residents,
- up to 90% reduction in crime,
- improvement in children's behaviour and improvements in the physical and mental health of residents. (Price Waterhouse 1997a: 2)

Architect Mary McKeown and the DICE team acknowledged these objectives but felt that they were not the drivers to the extent attributed in the report. She recalls that practical objectives (such as more efficient estate management resulting in fewer complaints or reducing repairs and maintenance costs) were discussed during selection visits, but felt these were not considered in subsequent analysis. Similarly, while these managerial benefits were important, Coleman argued that her interest was in wider social interaction, not solely reducing crime or anti-social behaviour. Sporle also recalled Coleman's expectations as more straightforward:

I didn't think Alice was claiming health improvements, I think she may well have said that it could make people feel better. But I don't think she was being scientific to the nth degree in terms of the actual impact. She dealt much more in generalities, the sort of common sense approach that she and Margaret Thatcher would have had, which is that if you've reduced crime then, yes, life is going to be better. (interview with Katrine Sporle, 2011)

There are always practical challenges in evaluating 'research in action' as there are in evaluating urban policy. Price Waterhouse had access to local crime data provided by the Metropolitan Police, but this proved problematic to interpret, with no consistent effects found across the five DICE estates. The most reliable crime reduction impacts were found on the Ranwell East and the Nottingham estates. This was similar for experienced crime incidents or witnessed crimes and incivilities. Fear of crime had also declined on the Rogers Estate but areal effects caused by changes in crime levels in the surrounding areas confused the picture. As such, the evaluation stated that the crime reduction analysis:

was least supportive of the Rogers Estate results, suggesting a particular local effect involving a general improvement of the whole locality. (Price Waterhouse 1997a: 69)

While the analysis for social control (which measured indicators such as ease of being able to identify strangers) revealed considerable variation with no clear pattern emerging, there was slight evidence of improvements in community control occurring on almost all of the estates. The exception, having a negative effect on levels of social control, was found on the Ranwell East Estate, suggesting that the design

changes there had increased permeability and numbers of non-residents accessing the site (Price Waterhouse 1997a: 71). Crime levels may have dropped on the
estates, but it was unclear if this was because crime had also reduced in the surrounding neighbourhood or because residents were able to make their estates safer.

This inconclusive pattern of findings and the struggle to interpret localised
crime levels is not unusual. There is a widely held view that crime can only be
managed, not reduced permanently (Schneider and Kitchen 2007; SNU 1993).
There is a need for repeated interventions over long timeframes to achieve a
sustained impact, as the case study of the Mozart Estate shows. Favouring Sampson et al.'s (1997) model of community collective efficacy as an explanation for
the slow process of building informal community control, Paul Wiles described
his experience with crime reduction projects:

> A crime reduction campaign may be unsuccessfully implemented on an estate, fail-
> ing immediately. However, a repeat project in the same location a couple of years
> later, might succeed as residents build on their earlier experience of failure and
> learn how to exploit repeated interventions. (interview with Paul Wiles, 2011)

Alternatively, Wiles described schemes where a significant crime reduction was
measured immediately after a physical intervention:

> This can be attributed not to any alteration in the building other than the process
> of decanting the resident population, disturbing the (anti)social networks before
> reintroducing tenants. (interview with Paul Wiles, 2011)

A frequent criticism of regeneration projects is that interventions have only
short-term effects and do not lead to long-term permanent improvement or
sustained impact (Palfrey et al. 2012), as illustrated in the case study of the
Mozart Estate (see case study following Chapter 4). SNU (1993) provided six
reasons for the transient nature of any changes:

- a time-limited project will have targeted, fixed resources that may be withdrawn;
- a very local project may have little influence on wider socio-economic
  circumstances;
- local interventions are not always able to influence the way borough-level or
  national agencies deliver services;
- a project might receive special services (such as additional neighbourhood
  policing) which later return to normal levels;
- successful but novel methods are not always disseminated within organisations
  beyond the initial project with transfer of positive lessons affected by staff
  turnover; and
- projects aiming to empower and educate local people often only minimally
  engage with residents and hence achieve less-sustained impacts than hoped.

The Price Waterhouse evaluative report failed to account for these wider explanatory factors, overlooking operational changes in Local Authority services, such as reduced grounds maintenance, frequency of police patrols or altered allocation/tenure policies. It also ignored broader economic and social changes, such as the 'right-to-buy' policy. A re-evaluation a year after the completion of the building works would have picked up any immediate and significant outcomes. But gathering a more representative, informative picture of sustained impacts would ideally require a delay of at least 36 months or more to indicate any permanence (SNU 1993: 157). In short, an evaluation to decide if an intervention is either a failure, or a (permanent) success, should not be limited to a snapshot at any one point in time.

Both monetary and non-monetary aspects were assessed in Price Waterhouse's evaluation, but despite the breadth of data gathered on non-monetary aspects (community instability, social behaviour, actual crime, fear of crime, tenant perceptions and satisfaction) greater prominence was given to the more easily measurable economic factors. The DICE programme was reported as costing somewhere between £43 m and £50 m for the research and the capital funding of the seven projects (*Building Design* 1988; Price Waterhouse 1997a). This £50 m covered capital funding for the seven projects, architects, consultants and Local Authorities' fees. Local Authorities' contributions varied, and paid for many of the new infill houses as well as additional non-DICE works such as new roofs.

Even the longer Price Waterhouse report (1997b) failed to communicate a rounded picture of the estates or the experiences of their inhabitants. Findings from hour-long, 40-page interviews with thousands of residents were compressed down to a mere six pages (interview with Toby Taper, 2013). But more critically, the report lacked the depth of material and basic facts that would be needed to persuade either policy makers or practitioners, enticing them to replicate the ideas:

> It was a fairly sketchy summary report. It doesn't stand up. It's not a piece of science. My test when I worked in government was that any research, any evaluation report that was published, that wasn't capable of being published in a major peer review scientific journal, shouldn't be published. Now I'm not sure that the final report for DICE would have met that standard, but I think it was because by then we were all fed up with the whole thing. I think we just agreed 'oh for God's sake let's just write a report'. (interview with Paul Wiles, 2011)

The Price Waterhouse report lacked a narrative that would intrigue and interest decision-makers to find out more. By way of contrast, the DICE consultancy leaflets were written as enticing, extremely positive case studies (Silver 1995, Coleman and Silver 1995a, 1995b).

Any evaluation needs to convince its intended audience that its recommendations have credible weight. In *Utopia on Trial*, despite using the rhetorical structure of accused, evidence and conclusion, Coleman's tone is personal and accusative; haphazardly targeting individuals, government departments and industry practice, or blaming the nature of bureaucracy itself. There is a flamboyant energy and bravado to her accusations, reminiscent of a barrister in eloquent persuasive flow. In comparison the Price Waterhouse report can be read as the defence response: taking a more impersonal, definitive tone and deferring to the department (DoE) as a voice of authority. The material is presented as self-evident, consensual truth. It is focused, responding only to points it can address and ignoring intriguing diversions and digressions. Much of this bland tone was intentional, to discredit Coleman's findings (interview with David Riley, 2011). The executive summary was a measured, ambivalent and carefully worded rejection of Coleman's methods and claims:

> None of the DICE schemes can be judged to have been effective in meeting the (admittedly ambitious) objectives set for it by Professor Coleman. Compared with the early Estates Action schemes (the most relevant policy alternative), the evaluation suggests that DICE was not a more successful regeneration initiative (nor at best does it appear to have been markedly less successful). (Price Waterhouse 1997a: 1)

Asked whether it was a difficult task to craft a credible, rigorous evaluation of concepts he had little faith in, Riley disagreed:

> The language was 'this was a really silly idea and we evaluated it thoroughly and showed that it was a silly idea'. I think the best thing was to have covered all the bases. The evaluation was 'respected', not the claims that were being made for intervention. It was brushed under the carpet. (interview with David Riley, 2011)

Alice Coleman's Own Evaluation of DICE

Alice Coleman viewed the Price Waterhouse evaluation as unscientific and intended to undermine her research. She had, perhaps not unsurprisingly, reached a state of exhaustion with the process and the DoE:

> They slowed it down. It was to have been a five-year project. In fact, it was nearly six, but they kept slowing it down and we never were able to look at these 70 other estates that we wanted to survey. Then they said, 'Yes, we'll give you some extra money for that … no, we won't …'. And it was just such a mess. Finally, I just wanted to get out of the whole thing. (interview with Alice Coleman, 2013)

Coleman's own reports state that the KCL DICE team was 'denied scope for systematic assessment of the results of our work' (Coleman and Cross 1994: 3).

Coleman was devastated that money for monitoring was unavailable from the DoE or Local Authorities. Yet despite this lack of funding the KCL DICE team undertook their own parallel evaluation of the project. They planned to map 'environmental abuse' scores four times: first, surveying the design features during the identification of potential sites, before residents on the estate were aware of the possibility of an improvement scheme. Second, once the estate had been selected and a conversation with the residents started, but before any works began. This second mapping was to test whether engaging with residents had any effect on levels of graffiti etc. Coleman claimed that she found very little change following the initial conversation with residents. Third, a mapping immediately after the refurbishment scheme finished. And fourth, a final survey a year after completion, to assess whether there was any lasting impact. However, this in-house evaluative plan was severely hampered by a lack of resources. The *Design Improvement Reports* for the Rogers and Ranwell East Estates show that both estates were assessed only three times. Although Coleman interviewed some residents informally to gather their views on the changes, these were not written up or systematically included in the final reports (Coleman and Cross 1994; Coleman et al. 1994).

From the first, Coleman's measurement was based on observations of the remnants of negative incidents. The KCL DICE team's mapping recorded the occurrence of litter, graffiti, urine and faeces. Vandalism was noted in 10 locations (fences, sheds, windows, doors, stairs, lifts, electrical fittings, refuse facilities, garages, building fabric) and scored as 'undamaged', 'damaged' or 'target not present'. The total abuse score was reported on a scale of 0–16 regardless of whether the 10 vandalism targets were present. Thus, a small walk-up block with no associated lifts, garages, sheds or fences would still be reported on the same 0–16 scale as a large multi-storey slab block raised above a row of disused garages with a far greater potential area for damage. Calling this process 'a mapping' overstates the geographical granularity achieved as, unlike in Wilson's vandalism research (Sturman and Wilson 1976; Wilson 1978b), the location of damage was not recorded. Other influential factors were overlooked too, such as whether damage occurred in an occupied space or in the proximity of empty flats. Wilson (1978b) found that ground-floor empty flats were most likely to be subject to vandalism. This means one occurrence of a damaged rain water pipe, a single broken light fitting and one broken garden gate would give the same DICE abuse score as a block where every door entry-phone had been systematically broken, several lifts had been maliciously damaged or with frequent instances of arson in refuse bins.

Crime levels distinguished by block would be essential to compare altered and unaltered blocks. This was a fundamentally different use of 'controls' to the Price Waterhouse approach who believed it was important to try to match both built form and social-economic factors within a tight geographical location, hopefully still identifying the fluid and mobile neighbourhood component of crimes and displacement. Coleman's inability to obtain crime statistics from the police was a serious difficulty and undermined her aspiration to replicate the detail of Newman's New York research.

Moreover, Coleman's evaluations were not made publicly available, which she felt demonstrated the DoE's dismissive attitude to the project:

> I think they did that [the Price Waterhouse evaluation] afterwards because we were supposed to be producing a report, which of course, we did, but they didn't want to publish it. They didn't want to know what we thought about it. They thought it would be more objective if they had their own thinking instead. (interview with Alice Coleman, 2013)

When considering Coleman's evaluation of her own project it is useful to try to distinguish the difference between research and evaluation (see Table 5.1), a distinction that is even more important in applied or action research like DICE. For some research and evaluation are seen as the same – they use the same methods, write similar kinds of reports and often come to the same conclusions (as evident in Coleman's research and evaluation of DICE); pragmatically evaluation can be seen as a form of applied research. But evaluation assesses the worth of a body of work, and programme evaluation like with PEP and EAP, and here DICE, asks practical questions and is undertaken in real life situations (in this case on council estates). The evaluations of PEP, EAP and DICE show that the boundaries between research and evaluation are actually rather fuzzy:

> Research and evaluation are not mutually exclusive binary oppositions, nor, in reality, are there differences between them. Their boundaries are permeable, similarities are often greater than differences and there is often overlap; indeed evaluative research and applied research often bring the two together. (Cohen et al. 2018: 81)

The design and methods in both forms of investigation are, after all, informed by social science methodologies and generate evidence. Indeed, it might be useful to think of research as a subset and tool of evaluation.

The SNU's *Crime Prevention on Council Estates* framework for evaluating the design interventions intended to reduce crime is precise about the distinct kinds of evidence required to demonstrate the impact of an initiative. Pre- and post-intervention assessment is needed against five categories of evidence: first, evidence that crime has actually reduced with confirmation that this decrease means a reduced problem (for example, altered severity overall or has crime been displaced). This is reliant on a mix of crime statistics and survey data as well as subjective assessment of actual occurrences or perceived fear of crime. Second, evidence of an initiative's effect and if it can be shown that the initiative was responsible for the change. Comparisons could be made against control areas, or sampling analysis used to check the consistency of the population under examination. Third, evidence of the impact and effect of individual measures.

Table 5.1   Research and evaluation.

Distinguishing feature	Research	Evaluation
Purpose	Generating new or expanding knowledge to inform subsequent research. 'Predicting what is and will happen'	Generating knowledge for programme or client The impact and effectiveness of intervention. 'What is valuable'
Decision-makers	The researchers	The stakeholders
Questions	Formulated by the researchers	Answers questions the programme is concerned with
Value judgements	Supposedly value neutral	Aims to create value judgement
Setting	A controlled environment	The action setting
Utility	Production, publication and dissemination of knowledge	Whether approach should be used in the future
Politics of the situation	Provides information for others to use	May be unable to stand outside the politics of the purposes and uses of (or participants in) an evaluation
Publication	Academic journals and other outlets mainly	Not always published, sometimes only stakeholders see reports
Use of results	Provides the basis for drawing conclusions. 'Theory dependent'	Provides the basis for decision-making, might be used to increase or withhold resources or to change practice
Use of theory	To create the research findings	'Field dependent' i.e. derived from participants, the project and stakeholders
Standards for judging quality	Judgements made by peers: standards include validity, reliability, accuracy, causality, generalisability, rigour	Judgements made by stakeholders also include: utility, feasibility, participation, efficacy, fitness for purpose

(Source: Adapted from Cohen et al. 2018)

Which of several measures has led to the change? When did the crime reduction take place? Four, evidence of permanence with follow up surveys to assess if the positive effects last? Fifth, evidence of replicability or generalisability. An objective, thorough evaluation should combine all of these five kinds of evidence to make sense of a complex social reality.

Considering these forms of evidence and the types of data it is clear that within an action setting there are many practical challenges to be overcome. The *Crime Prevention on Council Estates* report warned of the 'tensions between scientific and pragmatic approaches to evaluation' (SNU 1993: 15). The origin and focus of the investigation can increase these tensions. The key question is: Is it good research and worthwhile evaluation? To Pawson (2013: 190) evaluation can be prone to distortion and misrepresentation if it ignores the 'perceptions, actions and agency of those involved or it is selective on whose interpretation is promoted'. The rational expectation that a better evidence base will improve policy analysis obfuscates the less-rational world of diverse stakeholder values, ideologies and interests that socially construct that evidence base.

Comparing Price Waterhouse and Coleman's Evaluations

Both the Price Waterhouse evaluation and Alice Coleman's own evaluation of DICE tells us as much about the commissioning context and evaluative process as the epistemological differences between commercial consultants and academics. Comparing the costs of each evaluation explains the imbalance in the volume and type of data collection. The DICE evaluation was a prestigious and potentially lucrative commission for Price Waterhouse to have won, Sporle recalls:

> At the time, it was a big, big project. It was a million pound evaluation project. In those days that got you noticed by partners, working on a million pound project. (interview with Katrine Sporle, 2011)

The Permanent Under Secretary's undertaking to properly and fully evaluate DICE meant that the resources for the evaluation were extremely generous. Riley recalls the resources 'weren't artificially constrained, because he [Permanent Under Secretary Heiser[7]] was as interested in killing this off as anyone else' (interview with David Riley, 2011). Eventually the total payment to Price Waterhouse was over £1.5 M, with about £600,000 of this going to MORI for their resident surveys (interviews with John Harvey and Toby Taper, 2009). Significantly, both organisations were using consultants charging high rates in comparison to the cheaply paid university research staff at KCL.

Coleman was aghast at the amount paid to Price Waterhouse for the evaluation (interview with Alice Coleman, 2013). In comparison the research costs for
the KCL DICE unit were extremely modest. Yet the output from the KCL team
over five long years (40 initial reports, numerous site visits and at least seven
detailed evaluations) constituted a substantial body of work. The incompatibility
of 'commercial' consultancy evaluation with academic, theoretically 'objective',
research influenced all aspects of the two evaluations.

Coleman's long-term ambition was that the KCL DICE team would continue
after the DoE money ceased, funded via design consultancy activity. Her awareness of the potential brand strength of defensible space is visible soon after the
publication of *Utopia on Trial* with references to Coleman's *Design Disadvantage
in Housing Survey* almost as a trademark process (Bar-Hillel 1986b). Indeed,
the KCL DICE Unit is referred to as the DICE consultancy in press articles
from 1994 onwards. And, in fact, a number of Local Authorities and housing
associations did commission the KCL DICE Unit to provide design briefs for
estate improvement schemes, such as the consultancy work on the Mozart Estate
undertaken for Westminster City Council, suggesting that this might be the basis
for a workable business model. Of course, this was a period before research
enterprise divisions, when consultancy activity was still unusual in university
geography departments (even more so for human geography), which were under
no pressure to undertake external contracts. The impact and enterprise agendas
(part of the commodification of British universities) were not yet anticipated
and academic activity counted for more than external achievements. Coleman
was in a sense ahead of her time with her approach to gaining wider recognition
and influence via promoting DICE as a memorable brand, used in numerous
headlines such as 'An unlucky throw for the DICE projects' (Baillieu 1991b).[8]
Coleman's approach would fit very well into the current neoliberal university
'need' for large monetary research grants, impact and enterprise. Yet could Coleman really show impact in terms of improving council estates with defensible
space principles and alleviation of crime and anti-social behaviour? We argue that
this lasting impact was doubtful, difficult to prove and maybe even impossible
to prove. As the evaluations discussed here show, measuring (and thus proving)
impact is not easy!

The Price Waterhouse evaluation process was beset with difficulties from
the start. Unpopular from the outset with those who had commissioned it,
accused of lack of transparency or objectivity, the evaluation was impeded by the
fundamental challenges of urban policy evaluation. Located within an ontology
of hard 'scientific' assessment, it ambitiously attempted a quantitative assessment
of the physical, built environment. The evaluators were massively inundated by
the bewildering quantity of information gathered and confused by the growing
complexity of the analysis to try to address these methodological constraints. As
the costs of the evaluation grew, the staff resources to deliver it reduced. The final
drafting of the report passed from author to author as staff moved on to other

jobs. The Price Waterhouse report accused Coleman of ambitious aspirations for her principles, but fell into the same trap itself. A vast amount of data was gathered, but only a small proportion was analysed or released to a wider audience. David Riley and Paul Wiles stated that the slender provisional conclusions were the only ones that the team could deduce (interviews with Paul Wiles and David Riley, 2011), but even a cursory reading of the report cannot but notice omissions. One could ask, for example, who gained from such a slight discussion? In interview those involved in the report suggested two alternative explanations: that the evaluation had run its course, 'it just ran out of steam' (interview with David Riley, 2011), or a more premeditated political reason that the DoE 'smothered it to death by over-evaluating it' (interview with Katrine Sporle, 2011).

This kind of political obscuring or concealment is a frequent obstacle to the transfer of objective social science research into policy. Flyvbjerg (1998: 19) describes how in his study of the Aalborg project 'the evaluations became mere rationalisations of a political decision made in advance'. Despite endeavouring to professionalise evaluation processes, and the mass of theoretical literature justifying it as worthy of intellectual and academic concern, evaluation is still treated as 'a servant and not an equal of politicians' (Palfrey et al. 2012: 29). The Price Waterhouse evaluation was subject to the same pre-judgment that Flyvbjerg warned against, with the unspoken, but keenly felt political direction being to kill the DICE project. As Riley succinctly recalled, the aim of the evaluation was: 'puting the lid on the coffin and nailing it shut!' (interview with David Riley, 2011).

Yet, the cautious, almost tentative tone of the Price Waterhouse report failed to nail the coffin lid shut definitively. Even more cynically than Flyvbjerg, Suchman (1967: 168) identified a typology of 'pseudo-evaluations', which explain the failure of some evaluations in terms of having any lasting impact. He distinguished between superficial and shallow 'eyewash' evaluations from 'whitewash' ones intended to cover up programme failures, or those which are merely posturing lip-service or a diversion to postpone any practical action. His final typology (which it is not too harsh to suggest occurred with Price Waterhouse's evaluation of DICE) is the 'submarine' evaluation undertaken with the predetermined aim of undermining and sinking a project. The subdued launch of the Price Waterhouse report in 1997 indicated its irrelevancy to the DoE's opinion of the programme. As the civil servant responsible recalled, 'DICE was just a little bubble that got punctured' (interview with David Riley, 2011).

In comparison to the Price Waterhouse evaluation, the KCL DICE team's low-cost evaluation may seem lightweight, but as well as suffering from constrained resources it also suffered from methodological weaknesses, as well as a critical lack of access to crime data. The central survey technique in Coleman's evaluation relied on a visually subjective mapping of the occurrence of environmental 'abuses', which had already been roundly discredited. There was no discussion of the sensitivity of the abuse or design thresholds that the scores were derived from. With such approximate quantitative measures anything more than the basic analysis undertaken (weighted adjusted averages) would have been a spurious attempt at statistical accuracy. But a more fundamental flaw was the way in which the evaluation was approached – relying on a positivist scientific methodology to reveal facts. The KCL DICE team evaluation was setting out not to openly evaluate the success or impact of the DICE schemes, but to continue the process of 'proving' Coleman's theories.

Evaluation theorists such as Pawson and Tilly (1997) argue that current schools of 'realist evaluation' dismiss the possibility of theory-driven evaluations that presume to generate or confirm theories, except at an individual project or programme level. They claim the prospect of building a larger theory, that is applicable to multiple projects and could be used to assess large–scale national policy or public reforms, is unlikely (see Chapter 7 on defensible space as a middle-range theory). Yet perhaps government suppression of unfavourable outcomes on the one hand, and promotional simplification of evaluation on the other, should not be unexpected. The SNU *Crime Prevention on Council Estates* report concluded that the nature of evaluation is defined not just by the availability of data, but evaluators' professionalism and awareness of multiple interpretations:

> In fact, as a general rule, the *best* evaluated initiatives produced the worst results and vice versa. This in itself is not surprising. The more thorough evaluations are often carried out by researchers who, as we have said, tend to err on the side of caution. The least thorough evaluations tend to be carried out by practitioners, who may not have the time or inclination to search for further evidence and for whom least is often best. (SNU 1993: 99)

These two alternative evaluations of the DICE project from Price Waterhouse and Alice Coleman herself illustrate these circumstances well. One evaluation was overcome by excess data and analysis, the other limited by incomplete and sketchy data and the evaluator's one-dimensional intent. With these constraints any assessments/evaluation will always be partial and simplistic; yet often the harder you look for clear solutions, the murkier and more confused the picture may seem.

There was a muted reception to the Price Waterhouse evaluation and the DoE coasted through the mechanics of the report's launch. The report eventually appeared on the DoE website, but none of the interviewees remember any dissemination activities or publicity (interview with Paul Wiles, 2011). A defensive press release was prepared but was an unnecessary precaution as the press reception was subdued, due in no small part to the guarded way the report was written:

> We thought when it came out that she [Coleman] was going to explode and we were told 'Have a press release ready to say that it was all done through the most rigorous scientific method. We've got all the evidence, it's all been collected and collated and so forth'. But it just didn't get any coverage at all. I mean partly because there's

no highly quotable stuff in there. There's nothing that lends itself to that, is there? There's nothing that makes it tabloid or even *Inside Housing* kind of stuff. I think she was ill when it came out to tell you the truth. I think she just didn't have the energy or the wherewithal to do it and it just passed by. (interview with John Harvey, 2009)

The exact publication date of the evaluative report is unclear but Coleman recalls it as being early summer 1997 (interview with Alice Coleman, 2013). The press purdah period prior to the General Election in May that year would have restricted publication of DoE reports during the spring and as soon as the election had been announced there would have been fierce internal competition to prioritise those projects with the most political support. Post-election and following the switch from a Conservative to a New Labour government, the report might also have been tainted with negative connections to the previous regime. Surprisingly, only one article in the construction press during 1997 mentioned the evaluation. An article in *Building* in November 1997 featured a revisit to the Ranwell East Estate, evaluating the design changes and it reported the headline figures from the Price Waterhouse evaluation (Spring 1997). The housing press (*Inside Housing* included) and architectural journals, even those that had sustained interest in earlier disputes into the mid-1990s, were silent about the evaluation.

In private, Coleman was as displeased as Harvey had expected, which makes her public silence also all the more surprising. Having been such a dynamic force and so capable and tireless in her promotion of *Utopia on Trial* it seems odd that she made so little effort to promote or counter the evaluation. It would have been more in character for her to fearlessly and outspokenly defend her opinions, given she disagreed with the evaluation's findings.

In fact, Coleman's failure to comment on the evaluation is particularly unexpected as during this period she maintained a recognisable media profile being mentioned both in the mainstream and construction press. In 1996 and 1997 Coleman had a letter published in *The Times* (1996), was cited as a 'housing guru' in the *Evening Standard* (1997), and advised the London Planning Advisory Committee to increase residential density (*Building Design* 1997).

Peter Silver recalled that the timing of the launch coincided with two other events that greatly concerned Coleman: confirmation that the funding for the consultancy was not going to continue – 'the plug was to be pulled' – and that her position at KCL was ending. Coleman herself confirms this:

Nicolas Ridley, the Secretary of State, was pleased with DICE progress and said that after the five years he would like me to travel around the country to redesign as many other problem estates as possible. It seemed my retirement would be constructively spent to benefit numberless individuals, not only the estate tenants but also the victims elsewhere of the criminals they were breeding. This was probably my greatest high spot. It was not to be ... Heseltine became the Secretary of State and, as DICE was a Thatcher project, determined to foil it, supported by the civil servants who rejected Oscar Newman's work. He confiscated £10 million of DICE funding, replaced our helpful liaison officer with a hostile incompetent, had me report to someone who told me not to solve problems and banned our scientific comparison of improved and unimproved blocks with the same disadvantagement scores. Instead, he had civil servants write up a condemnatory report on DICE. (interview with Alice Coleman, 2008)

On her retirement from the Geography Department at KCL in 1995 Coleman's intention had been to continue to work on the DICE research in an emeritus position with a desk at KCL for a further five years. This reduced to two years and with working conditions at KCL consisting of a small desk in a shared room, Coleman found it easier to move back home, drop the 'Thatcher project' and to turn her attention to other projects (interview with Alice Coleman, 2013). Despite Coleman's advancing age during the period of the project (she was 61 when *Utopia on Trial* was launched and had reached her 70s when the evaluation was finally published), colleagues described her as full of energy and enthusiasm. Indeed, when Loretta Lees started work in the Geography Department at KCL in 1997, Alice Coleman would still come in to collect her mail and chat to people, including introducing herself to new lecturers, like Lees, who had just returned to the United Kingdom after two-and-a-half years in Canada.

It is clear that the evaluations of DICE did not definitively characterise it as a policy success or failure. In fact, it was situated somewhere in-between. McCann and Ward (2013: 2) have identified policy success and failure as impermanent states, like policy mobility and immobility, not dualisms but interrelated tensions in policy mobilities research. They argue:

Neither success nor failure is absolute. One does not make sense without the other. Rather, success and failure are relationally constituted in politics and in policy-making.

To a degree they have a point, but leaving outcomes in dialectical play is perhaps the easy way out. In Lovell's (2017) discussion of the dearth of work theorising policy failure she argues that policy mobilities research is missing large parts of a wider empirical picture. This chapter (and those before it) has looked at that wider empirical picture and demonstrated the role of politics and power in constructing success and/or failure. Defensible space in these terms can be seen as a contested practice on council estates. Price Waterhouse saw DICE as a 'site of failure' (Jacobs 2012). Coleman saw it as a 'site of success'. But success or failure seems almost irrelevant given the resilience and continued mobility and mobilisation of defensible space, and mobility itself is not easily connected to success or failure. The relationship is ambiguous and the assemblage is fragile and can break down at any moment. The success and failure of defensible space looks different with respect to different disciplines, e.g. geography, criminology and architecture. Success or failure can be political (here defensible space failed because of its association with Thatcher and political support dropped off), programmatic (here, as we said at the beginning of the chapter, the DICE programme was in competition with EAP), or process based (the research design was flawed) (see Marsh and McConnell 2010, on this classification).

The absence of discussion after Price Waterhouse published their evaluation cannot be ignored either (on absences see Müller 2015; Webber 2015; Lovell 2017). The emergence of this absence was built into the writing of a dull and limited report that got little attention from the usual suspects, including Coleman herself. Whereas success is always contested (Marsh and McConnell 2010: 575), so too is failure. Defensible space from its conceptualisation and then implementation through DICE was never positive or negative, proven or unproven, a success or a failure; it was translated and codified through the contestation that took place within these dualisms. The mobilisation of defensible space was strongest during the Thatcher endorsed 'policy window', yet, despite the negative evaluation of DICE by Price Waterhouse, the concept and approach remained resilient and its uptake continued, as we discuss in Chapter 6.

Notes

1 The *British Crime Survey* [BCS] (1984) identified three high-risk residential areas: 1) inner-city high-status non-family areas (rich homes/privately owned buildings in multiple occupation); 2) inner-city multi-racial areas; and 3) the poorest council estates (*BCS*, 1984, cited in SNU 1993: 8)
2 The principles underlying the local management are summarised in the DoE Housing Services Advisory group (of which Sheena Wilson was a member) report *Security on Council Estates* (DoE 1981). This emphasised the need to address fear of crime and the feeling of social isolation arising from tenants' lack of belonging, as well as insensitive or remote housing management.
3 In 1994 Katrine Sporle CBE became Chief Executive at Basingstoke and Deane Borough Council, before becoming Chief Executive of the Planning Inspectorate for England and Wales (2003–2011) and Property Ombudsman in 2015.
4 Tim Hope was also an evaluator of the PEP (see Foster and Hope 1993).
5 *The Magenta Book*. https://dera.ioe.ac.uk/10521/1/complete_Magenta_tcm6-8611.pdf.
6 The five of the seven DICE estates evaluated were: Rogers and Ranwell East in Tower Hamlets; Kingsthorpe Close, Nottingham; Avenham Estate, Preston; and Bennett Street, Manchester. Excluded were, Nazareth in Birmingham and Durham in Sandwell, seemingly for no clear reason other than cost and programming.
7 Sir Terence M. Heister, DoE Permanent Under-Secretary, 1985–1992.

8   This inward focus was on the point of change as the parallel experience of Space Syntax shows. Their design assessment of the Mozart Estate was undertaken by Hillier and Penn in 1986 when they were part of the (still academic) Unit for Architectural Studies, but three years later in July 1989 UCL established Space Syntax Ltd as a separate commercial entity. It is now a very successful global consultancy with registered offices in a dozen countries across Europe, Asia, the Americas and Australia.

Chapter Six
The Uptake and Resilience of Defensible Space Ideas

It is also simply easier to commit crime in the complicated concrete and glass jungles of modern multi-storey housing. It offers a plethora of semi-private, semi-public unpoliceable spaces such as corridors and stairwells which are hard to survey and which offer multiple escape routes. (Boys Smith and Morton 2013: 35)

In 1985, Alice Coleman, a fierce University of London geography professor and disciple of Newman's who became a policy guru for prime minister Margaret Thatcher, used the Aylesbury as a case study in an unforgiving survey of postwar council estates, *Utopia on Trial*. Her and Newman's claims about the malign effect of communal areas and high-rise living on the behaviour of residents were criticised by many academics as too deterministic, and for largely ignoring the impact on estates of the wider economic and social upheavals of the 70s and 80s. The ideas still seeped into national and local government. To this day, you frequently hear Newman's 44-year-old arguments – that private space is better than public space; that public space must be overlooked by homes to be safe – repeated verbatim by Conservative ministers and Southwark councillors, but unattributed, as if these notions were mere common sense. (*The Guardian*, 2016)[1]

Despite the problematic evaluations of defensible space, and inconclusive proof of its utility, the concept has proven to be resilient. During the protracted 2018 Aylesbury Estate Public Inquiry, where objectors sought to halt the demolition of one of Europe's largest (modernist) public-housing estates, the lack of defensible

Defensible Space on the Move: Mobilisation in English Housing Policy and Practice, First Edition. Loretta Lees and Elanor Warwick.
© 2022 Royal Geographical Society (with the Institute of British Geographers). Published 2022 by John Wiley & Sons Ltd.

space that Oscar Newman had highlighted on the Aylesbury Estate (see Chapter 2) was advanced as 'evidence' that the estate needed to be demolished. This repetition of Oscar Newman's 1974 criticism was deliberate, with the You-Tube video of Oscar Newman on the Aylesbury Estate[2] mentioned in evidence, having been watched by the barrister for the objectors (see Lees and Hubbard 2020). Stephen Platts, Director of Regeneration for Southwark Council, in giving evidence for the demolition, claimed designed-in 'blind spots' on the Aylesbury Estate 'had increased criminal behaviour' and 'the monotonous blocks made finding one's way around hard'; saying 'it doesn't feel like a normal London neighbourhood'.[3] But later in the public inquiry others reproduced the more confused stance that can often be found in discussions of defensible space. Simon Bayliss (who led the multidisciplinary design team for the regeneration of the Aylesbury Estate) said:

> I think Alice Coleman would prefer us all to live in terraced houses with a garden – she lives in Dulwich, I disagree with her![4]

Despite saying this, Simon Bayliss went on to promote his new regeneration master plan as one that would design-out crime from the Aylesbury Estate. Indeed, Southwark Council spent some time trying to prove that the historic design of the Aylesbury Estate had caused what its Senior Design and Technical Officer, Catherine Bates, called a 'very high fear of crime'.[5] She said they needed to take down the walkways and bridge links to address crime and anti-social behaviour – in a similar vein to Coleman (1985a) – but provided no evidence to support the claim. Michael Leary-Owhin, the lecturer in urban planning and regeneration employed by Southwark Council to counter those academics presenting evidence for the objectors, acknowledged that: 'Alice Coleman was extremely influential on government policy',[6] whilst also providing no evidence on the nature of this impact or any subsequent uncertainties around portraying geographer Alice Coleman as a kind of policy guru.

This use of Newman's and Coleman's canon is notable for the way that over 30 years later, defensible space ideas are still being used to put British modernist, high-rise council estates on trial, here at a Public Inquiry that Leary-Owhin (2018: 3) considered 'one of the most important ever in the UK'. In this chapter we use the concept of resilience to explain the continuity of ideas on defensible space in English housing policy and practice. Part of this continuity is about change of a concept to fit different circumstances with the attendant (neoliberal) agents of this continuity or change. By resilience we mean the ability of an idea to continue in the face of internal weaknesses and external challenges (cr. Schmidt and Thatcher 2013). We have described how many evaluations of the concept of defensible space, for the most part, repudiated it; and how its utility has remained ambiguous (neither proven nor disproven). To explain this resilience, we look now at how defensible space was developed, adopted, spread and maintained in subsequent years. This is an ideational resilience: communicated, generated and carried by individuals and discursive communities. There are of course processes and mechanisms that promote resilience, as well as the feedback effects that serve defensible space through supportive and reinforcing processes. Following Schmidt and Thatcher (2013), but also policy mobilities work (e.g. Stone 2016), we explore the continuity of defensible space through notions of bricolage (new ideas grafted on), conversion (the idea of defensible space used in new ways), diffusion (spread) and translation (adaptation to new contexts). To do this we move the focus away from individual influential transfer agents to tracking the threads of defensible space through government policies, initiatives, commissions, inquiries etc. Individuals still play an important role, but this shift is indicative that despite this constant movement and evolution defensible space was firmly embedded in English housing policy.

This chapter describes how the rather abstract concept of defensible space managed to jump the gap between theoretical and practical knowledge and successfully embed itself into British construction industry guidance. As we have shown earlier, the policy context for built-environment crime prevention during the DICE period was shaped by growing inter-departmental exchange, realigning crime prevention from the HO's detection of offenders towards the DoE's preventative deterrent as an outcome of housing/planning policy. Yet the primary policy instrument for embedding crime reduction in housing, the SBD initiative, emerged from the HO's Crime Prevention Unit and it was the HO that initiated the longest lasting form of 'pseudo-policy', which still remains the most current codification of defensible space – British Standard (BS) 8220 *Guide for Security of Buildings against Crime*.

The British Standards Institute (BSI) is an independent national standards body, so while not a formal governmental policy-making agency, BSI nonetheless exerts considerable influence on government thinking as well as shaping industry activities. The BSI forms a normative corpus of quality standards for products or services, referred to by the majority of architects using the National Building Specification and thus can be considered a highly familiar form of policy advice. Following the inner-city riots in English cities in the early 1980s, the HO asked the BSI to form a new committee to write a standard for securing homes against crime (Anson 1986). As justification, it cited HO crime statistics on homes as most subject to burglaries, criminal damage and vandalism (BSI 1986). Given the extensive debate over her research, it was unsurprising that the committee considered Coleman's findings.[7] In the conclusion of *Utopia on Trial*, Coleman positively reports the inclusion of 10 of her 12 design recommendations for houses within BS *8220-1:1986 Guide for Security of Buildings against Crime* (Coleman, 1990 rev. ed.). This overstated the case. A close reading of BS:8220 finds some general points corresponding to the DICE principles and only selected principles for houses and communal blocks, without the quantified parameters dictated by Coleman.

The intended audience for BS:8220 was unusually broad for a technical document, targeting residents as well as architects, builders and crime specialists. Despite this accessibility, the guide emphasised the need to consult experts, including insurers, locksmiths, fire prevention officers or police officers with crime-prevention experience (this was just prior to the formal designation of Architectural Liaison Officers within police forces). It stressed that judgment was required to interpret the guidance and to assess acceptable levels of risk against cost effectiveness and whether the measures might impact negatively on residents' quality of life. Balancing all these concerns made a simple generic solution unlikely:

> Often there will not be a single clear cut answer to a security problem, because many of the factors involved in the assessment of the risk are themselves known imprecisely. (BSI 1986: 1)

Local contextualisation of advice was critical, with all 'measures chosen according to local circumstances' (BSI 1986: 4).

As well as guidance on the layout of new estates and multiple dwelling blocks (such as avoiding communal entrances concealed from public view), the BSI advised on improving the security of existing estates. The standard endorsed the general principle of defensible space providing territorial control for residents:

> as a general principle space in communal areas, whether inside blocks of flats, or around groups of dwellings, should have its purpose defined and allocated, so that residents may supervise and exercise some control over their environment. (BSI 1986: 4)

It reiterated that physical security measures alone were an ineffective solution on existing estates and that good locally based management, improved staffing and maintenance were an essential component of estate security. BS:8220 was concerned with construction and product specification, the fixing of doors, locks and security shutters, but critically also emphasised the detailed design of elements that the DICE principles failed to mention such as location of car parking, external lighting (particularly to entrances, garages and communal areas), access controls or entry phones, vandalism targets other than lifts or door closures, rainwater pipes, fences, sheds that might aid unauthorised access to windows and the detailed design of open spaces, particularly planting and fencing to ensure enclosure.

Secured by Design

As defensible space ideas filtered into British policy-making circles, SBD, which is the British version of CPTED [*Crime Prevention Through Environmental Design*], came into being under the auspices of the Association of Chief Police Officers. Like CPTED, SBD, which started life in 1989, led to police officers being trained as crime-prevention design advisors, known as Architectural Liaison Officers, and has created design standards and an awards scheme. Its influence is considerable: planning permission for all public buildings, housing and schools is now contingent on meeting SBD standards.[8] (Minton 2013: 6)

Although robbery and violent attacks on individuals often have an environmental component, both the international initiative CPTED and the British SBD concentrate on burglary, vandalism and crimes against property and promote both physical solutions such as target hardening (fitting locks, alley gating) and neighbourhood watch schemes. CPTED is a 'multidisciplinary approach to deterring criminal behaviour linked to reducing opportunities for crime through environmental design, from structures to neighbourhoods' (Gamman and Thorpe 2014: 6). Chapter 1 explained how the origins of CPTED emerged from the writings of Jane Jacobs (1961) and Oscar Newman (1972) but Armitage and Ekblom's (2019) more recent work is an attempt to update CPTED in line with developments in design, architecture and crime science. Current criminological debate identifies the contradictory role of strangers or familiar neighbours (Armitage and Ekblom, 2019; Cozens 2015; Ekblom 2011; Reynald and Elffers 2009). Guardianship of spaces and manageability is related to the numbers of people sharing a space, whether they are known, plus mechanisms for reinforcing the social bonds needed to establish positive territoriality. From this, the second generation CPTED explores concepts of community, neighbourhood capacity and cohesion; ways of designing in the connectivity leading to occupancy, intensive use and visibility that overcomes the opportunistic accessibility and privacy needed for criminal activity.

As urban design debates centring on the benefits of connectivity and permeability of the public and private realm have redoubled, criticisms of CPTED's exclusionary characteristics have increased (Cozens and Love 2017), particularly in situations where excess target hardening results in unnecessarily high levels of security and restricted access. Picking up on Coleman's conviction that the symbolic design of entrances could be a sufficient deterrent to unwanted access, Cozens (2016: 13) questions the consequences of CPTED's intrinsic intention to exclude offenders/criminals and reliance on target hardening, asking whether 'the reduction in quality of life from the target hardening is greater than the benefits in crime reduction?' He quotes Raymen's (2015: 2) concern that following CPTED principles unthinkingly is 'designing-in the decline of symbolic efficiency and the development of potentially harmful subjectivities by designing-out the social'.

This conflict between place design and crime reduction principles is evident in the spilt government responsibility for residential crime advice, which moved back and forth between the HO and the DoE during the 1980s. BS:8220 encapsulated the DoE's advice on defensible space to the construction industry, providing technical background for the DICE and other regeneration schemes of the era. As we describe later, the 1994 edition of BS:8220 incorporated many of Coleman's DICE principles until it was heavily revised in 2000, reverting to a less prescriptive version of defensible space. Yet it was the HO who eventually operationalised defensible space principles into a widely accessible format, SBD, stimulating a whole industry able to provide the technical expertise to deliver this design advice.

SBD began as a regional crime reduction initiative in the mid-1980s, in the form of a secure property labelling system for homes encouraging the building industry to design-out crime at the earliest planning stage. Initially funded by the HO and administered by the Association of Chief Police Officers (ACPO), it was delivered on the ground by a network of local police known as Architectural Liaison Officers (ALO), Crime Prevention Design Advisors (CPDA) or Designing-Out Crime Officers (DOCO) located in local police services or planning authorities. New housing developments following the guidance were awarded SBD approval and allowed to use the logo (see Figure 6.1) in promotional literature. SBD became an independent private company in 2000 – supposedly to relieve itself from political interference (Minton 2013) – funded by the 480 or so security companies selling security products (door locks, alarms, windows etc.) certified to SBD standards, with profits being used to pay for ALOs. While ceasing to be a mandatory regulatory requirement in 2015, SBD still retains its status as a frequently used, well-respected tool by architects, planners, builders and housing managers.

In terms of achieving a rapid dissemination the wide application of SBD can be considered a successful policy initiative. The SNU report (1993) cites SBD as an example of how widespread and accepted design-led crime-reduction interventions had become. By 1994 almost all British police forces had ALOs

Police Preferred Specification

Figure 6.1   The Secured by Design logo.

or CPDAs. A HO manual described their role as reinforcing the physical environment's significant influence on criminal behaviour, via an offender's reliance on opportunity, anonymity, easy access and quick escape routes (Home Office Crime Prevention Centre 1994). The DoE circular 5/94 published in the same year was the first guidance that cited crime prevention as a material consideration in determining planning applications. The circular set out four principles for designing out crime: considering crime prevention from the outset; providing mixed uses while avoiding those that could cause conflict; careful layout design to reduce the risk of criminal activity; and encouraging consultation with trained specialists – the ALOs (DoE 1994). The concept of police providing advice on housing design was not novel. In fact, as early as 1985, Coleman had recommended to the Commissioner of the Metropolitan Police, Sir Kenneth Newman, that police officers should give design advice to planners and housing developers to aid crime prevention:

> The second practical application of the research results would be the introduction of a right for the police to vet plans for new housing similar to that exercised by the Fire Brigade. Unlike fire vetting, however, crime vetting could be exercised in a preventative sense. Objections could be raised to any and every design variable that breaches its threshold, so that disadvantagement scores are kept down to zero, at a level where crime is absent or very low. (Coleman and Brown 1985: 54)

On the basis of her report and this recommendation, Sir Kenneth Newman asked Coleman to provide training on mitigating crime through design to a group of police officers who would become an early version of the ALOs. Coleman gave 28 training sessions across England and Scotland, but recalled 'the moment Thatcher had gone, they stopped. It was a pity because they [the police officers] quite liked them' (interview with Alice Coleman, 2013). So Coleman's operational version of defensible space was a foundational premise for the establishment of SBD and the ALO taskforce.

Similarly, CPTED's origins can be clearly traced from Newman's version of defensible space (Ekblom 2011; Jeffrey 1999), and Newman's principles also underpin SBD (Armitage 2000). The five main themes of the SBD award scheme are: physical security, surveillance, territoriality, access/egress and ongoing management. SBD's primary focus is on the physical security of individual properties and less at the scale of the design of estates. The second and third themes more closely follow Newman: aiming to achieve social control by optimising natural surveillance and territoriality, while maximising private and minimising public space (Armitage 2013). The final two themes reiterate Coleman's concerns: limiting the number of access points onto estates; deterring non-residents and potential offenders; and requiring a programmed management system to maintain a clean and orderly environment as a signal to offenders that crime will be noticed and not tolerated.

SBD was originally devised for new dwellings and has been updated several times since 1989, with the latest version published in 2019.[9] The first explicit guidance for refurbishment was a scant page-and-a-half published in 2004. This extremely high-level summary acknowledged the difficulty of retrofitting the SBD principles to existing buildings, estates or listed buildings. The final paragraph (quoted in full here) describes applying SBD to major refurbishment projects (such as DICE on Rogers, Ranwell East, or the Mozart Estates):

> The involvement of existing residents should feature in the implementation of SBD guidelines. The residents will have first-hand experience of the crime risks and the practicality of any proposed security improvements. Also, their approval and co-operation is crucial in ensuring newly installed security hardware is properly used when the project is completed. Where such property improvements involve central government funding, the ALO's report for SBD approval (which may include a crime profile of the area) can be used to support a local authority bid for resources. (ACPO and Secured by Design 2004: 2)

Regardless of this brevity SBD was, and still is, universally referred to in refurbishment specifications.[10] It defined the required criteria for the *Form B* submission to the DoE, and also for the application for central government regeneration funding for the Mozart Estate described in the case study in this book.

Several studies assessing the effectiveness of SBD (Armitage 2000, 2004; Armitage and Monchuk 2011; Brown 1999; Pascoe, 1999) concluded that the scheme provided a cost-effective (or at least cost-neutral) way to reduce crime in housing. Yet Cozens et al. (2001a) argued that the advice SBD practitioners were passing on was based on ambiguous evidence. An early evaluation of SBD by the Building Research Establishment found that it was already highly proscribed, recommending: 'returning to the greater flexibility shown by some of the designers at the creation of the scheme' (Pasco 1992: 114). This detailed critique of SBD mentioned several issues familiar from the critiques of DICE, including: the fixed format of SBD; that SBD's foundational rationality assumed universal behaviour in specific places despite the unpredictability of individuals; that SBD paid little attention to who lived in the housing development, needing greater consideration of socio-economic context; and that there were fundamental uncertainties about the criminal or social problems SBD was trying to address – for example whether it targeted preventing burglary, auto theft or youths hanging around estates. Like DICE, SBD attempted to address too diverse a breadth of problems. Pasco (1992) questioned the theoretical basis of SBD, as recent criminological models of offender perspectives differed substantially from the SBD fixation on simple layout and target-hardening measures. Planners were accused of endorsing a simplistic concept of SBD as cul-de-sacs surrounded by high fences. Practically there was uncertainty over the target 'customers' for SBD: planners, developers, purchasers, tenants, or their neighbours; each of these groups wanting distinct outcomes rather than a generic desire for 'feeling safer'. The timing of an ALO's engagement in providing advice was found to be critical and, significantly, the inconsistency in the interpretation and application of SBD principles resulted in an uncertain appraisal process. An unpredictable appraisal process may, of course, result from erroneous formative principles or more practically, poor training of those implementing it.

Despite her involvement in the training of the early version of ALOs, Coleman disapproved of SBD as it emerged. In the second edition of *Utopia on Trial*, Coleman's criticism of the DoE was transferred to the HO:

> When *Utopia on Trial* was written the DoE was the villain of the piece; its powerful bureaucracy had subsidized and dragooned public housing into a monolithic Utopian conformity which it refused to recognise as a disaster. Secretary of State Nicolas Ridley changed that when he set up the DICE project specifically to undo some of the damage perpetrated by the DoE's design misguides ... Just as the DoE is moving away from counter-productive design dictatorship, the Home Office has stepped in to reinforce it with an advisory video and seal of approval system [Secured by Design] urging all the design defects. These are to be offset by locks, bolts and other security devices. (Coleman 1990: 183)

In her opinion SBD 'shows up the things that cause crime and makes them secure by locking them in place rather than designing them out ... People may feel happier but they're just imprisoned' (Coleman, quoted in Baillieu 1991a: 24). In the same article Coleman argued that any resultant crime reduction would be temporary and, by following SBD, government money was being wasted displacing crime rather than curing it.

SBD was promoted as design for community safety, yet it downplayed the role of the community in creating a safe environment, ignoring the idea that active inhabited public places create natural surveillance and ownership. Target hardening and CCTV was seen as a panacea, placing too great a reliance on technology for surveillance. Other non-physical interventions such as shared maintenance, concierges or Neighbourhood Watch schemes could address these issues more successfully. Community architect Ben Derbyshire[11] concluded that the contribution of these social mechanisms outweighed those of physical design, particularly when remaking existing places:

> The basic rules of defensible space are robust and sensible to apply but it's not always possible to retrofit estates successfully. [One should] acknowledge that more often a sensitively managed relationship of residents and community with the estate managers is needed to overcome some shortcomings of the physical surroundings. (interview with Ben Derbyshire, 2012)

Ben Derbyshire's questioning of the primacy of design (whether for DICE or SBD) reinforces the breadth and flexibility of on-the-ground practitioners' solutions to community safety against Coleman's and others more simplistic approaches. The *Building* article (Spring 1997) on the DICE changes to the Ranwell East Estate extensively quotes Mark Jones who was then the Local Crime Prevention Officer for Tower Hamlets Council. He suggested that social mix and intergenerational conflict continued to remain a problem post-DICE despite improvements to the physical layout. The estate might be a nicer environment to live in, yet social problems persisted. Jones, who at that point had worked in crime prevention for 15 years, continued to work as a CPDA based in the Borough. Over this 30-year period his personal definition of defensible space evolved:

> Initially, for me, defensible space would have been very rigid; railings, tall fences, grilles on windows, barbed wire, perhaps, very naively thinking that big is beautiful and will solve a problem. Over the years, what's changed, partly because of working with planners or architects, is the realisation that you have to look at who's going to live there, what impact it has on them. It's becoming more socially aware. Now it's the smaller things like putting up a row of prickly bushes or a little knee-high fence to designate private, semi-private, semi-public space. (interview with Mark Jones, 2011)

Jones' critique of SBD echoes the academic ones of excessive fixity and overall rigidity. He notes that he was prepared to compromise or to depart from the official line on elements where his experience had taught him an alternative response was workable:

> If you showed me an estate 15 years ago, and said what's your opinion? I would have said do this, do this. Now I would say well, you could do this and this, but you could also do this, this and this ... *Secured by Design* is like that as well, it can be very black and white. There are certain things I won't compromise on, but most of the things in *Secured by Design* are up for compromise ... It has to be local. You can't legislate in any way, shape or form, for everything. And you need to be able to say to the local representative, whether it's a planner or a *Secured by Design* officer or a builder, you're local, you know the area, what is appropriate for you? We've got the principles, but we need to be able to adjust and change depending on where that estate is. (interview with Mark Jones, 2011)

Jones' perception is that sites that incorporate SBD principles generally have lower crime rates than non-SBD sites, but this is based on experience not empirical evidence. His own analysis of sites he has advised on suggested a 5% reduction in crime, which he acknowledges is a small improvement, but sufficient to make a difference to residents' lives:

But we just know in our hearts and our heads that *Secured by Design* does work, even if it's just putting the right doors and glass in and nothing else. (interview with Mark Jones, 2011)

The crime-reduction and urban policy context also evolved with the rise of community-centred over design-led intervention strategies (McLaughlin and Muncie 1996). This filtered through into the technical guidance. In 2000 a dramatically rewritten *BS 8220-1:2000 Guide for Security of Buildings against Crime: Dwellings* was published, again at the request of the HO. The revised BS:8220 opened with the following realistic, if pessimistic, statement:

> Crime problems are caused by a multiplicity of factors, family, social groups, education, moral culture, drug and alcohol abuse etc. The influence of the layout of neighbourhoods and the design of buildings on these factors is limited. (BSI 2000: 1)

This indicates a swing away from physical determinism, returning towards the HO's model of crime being caused by opportunistic individuals, with the built environment merely influencing guardians, victims and offenders' behaviour. Illustrating the growing professionalisation of crime prevention in the intervening 14 years, residents were no longer an intended audience, with BS:8220 reiterating that the standard should not be regarded as a substitute for expert advice (BSI 2000). Design tactics such as small, defined clusters of homes were expected to maximise territoriality to such a tangible extent that occupants would be encouraged to challenge potential offenders:

> One common method of achieving this [territoriality] is by the application of 'defensible space' concepts i.e. classifying space into four different kinds of spatial areas, public space, semi-public space, semi-private space and private space. (BSI 2000: 4)

Reiterating Newman's original typologies of space, it highlighted the blurred boundaries between semi-public/private spaces in multi-occupier developments. The BSI warned against offenders' practised familiarity assessing the deterrent quality not only of physical barriers, such as fences and gates, but also of symbolic ones (signage, archways, textured surfaces). A section titled *Defensible Space for Ground Floor Dwellings* was limited to the use of plants to screen windows without blocking natural surveillance, with no indication of how to achieve these, sometimes contradictory, design requirements in practice. Nonetheless, the overall message at the turn of the Millennium was that design – either of dwellings, the layout or of the security products specified – was not a sufficient solution on its own.

Defensible Space Post-DICE

Post-DICE and under a New Labour Government, planning policy started to take a more holistic assessment of what made a good place and what was required to create 'safe, sustainable, liveable and mixed communities' (ODPM 2005: para 5). Policy, which had until then focused on housing provision as a separate entity, started to portray housing as a way to establish communities and social cohesion (see Imrie and Raco 2003). Kitchen and Schneider (2005) attribute this broadening interconnectedness of crime to other urban policy issues as a response to the perceived failure of more traditional and isolated approaches to crime reduction. They describe the extensive policy agenda and 'plethora of initiatives at both national and local level' (Kitchen and Schneider 2005: 276) but with politics and 'political reaction' distorting the interaction of the factors (cr. Flyvbjerg 1998, 2001). Historically, crime has long been used to justify government intervention in actions shaping housing: from slum clearance, allocation of homes, to the promotion of homeownership. Equally, each of these is indicative of general political views on the purpose and value of social housing and more generally of public services. Madden and Marcuse (2016) argue that the fundamental political underpinning of government intervention in housing problems is merely an example of the state preserving political stability and continuing to support 'accumulation of private profit' under the guise of urban renewal. Thus, the political distortion of housing policy is unavoidable:

> The very term 'housing policy' is evidence of this myth. The phrase itself suggests the existence of consistent governmental efforts to solve housing problems. But a historical analysis of governmental actions and in-actions affecting housing reveal nothing of the sort. Housing policy is an ideological artefact, not a real category. It is an artificially clear picture of what the state actually does in myriad uncoordinated and at times contradictory ways. (Madden and Marcuse 2016: 119)

The first London Plan in 2004 demonstrated this politicised overlay of interconnected and conflicting initiatives. Housing policy was just one strand of a strategic design-led planning vison encompassing all scales: from the capital's growth agenda, to the sustainability, connectivity, accessibility and safety of individual public spaces (Carmona 2012). Increasingly highlighting defensible space and crime reduction as essential actions for delivering popular residential environments and desirable prosperous public realms, it reinforced the state's movement towards detailed planning controls and the protection of property rights. Subsequent iterations of the London Plan under differing political regimes repeated these concepts: for example, higher density housing surrounded by controlled overseen spaces, merely reallocating roles and responsibilities for greater surveillance to validate privatisation and exclusionary management of formerly public spaces. In practice, regardless of the political motivations, the 'new' urban renewals mooted often resulted in the same unhappy consequences for the communities priced out and displaced from their existing neighbourhoods (see Hubbard and Lees 2018; Lees 2014).

The Urban Task Force, riding the political tide of Blair's election in 1997, was convened by the Department for Environment, Transport and the Regions (DETR) to devise solutions to reverse the decline of inner cities and the impact of growing suburban sprawl (Lees 2003). Chaired by the architect (now Labour Peer) Richard Rogers, its proposals called for greater investment, innovation and engagement at an urban and political scale (see Imrie and Raco 2003; Rogers and Power 2000). One of only three women in a Task Force of 14, Anne Power brought to it her understanding of housing, community-based neighbourhood management and the particular regeneration needs of council housing areas. Despite the idea of carefully increased areas of mixed-use intensification, accompanied by more localised hands-on management, higher density was crudely interpreted as making high-rise living acceptable again. As a result, in 2000 and 2005 central government's *Planning Policy Guidance Note 3* was deliberately changed to direct planning authorities to intensify inner city sites encouraging higher density, high-rise developments. In London, the then Mayor, Ken Livingstone, was also a fan of high buildings and he fed this into The London Plan: 'the Mayor will promote the development of tall buildings … [the boroughs] should not impose unsubstantiated borough-wide height restrictions' (GLA 2004: 181). This was caveated by boroughs being able to set locally based height thresholds in their Unitary Development Plans, but the pressure placed on planning authorities to ignore these was intensified by London's subsequent Mayor, Boris Johnson, after 2008. This set the scene for the high-rise densification that is continuing in London (see Blanc et al. 2020) and many cities across England today.

One example of this new, high-density London residential vernacular is the new Kidbrooke Village development which has replaced the Ferrier Estate in Greenwich. As Boys Smith and Morton (2013) note:

> Kidbrooke Village has no concrete panels, a better name and is being well advertised. But little else has changed. In the last analysis, we have replaced one enormous development of tower and slab-blocks with few real streets with another enormous development of tower and slab blocks with few real streets.

Boys Smith and Morton (2013) see the newly built Kidbrooke Village as very similar in design to its modernist, high-rise predecessor, the Ferrier Estate, implying a continuance of the problems that went with that old design. The new masterplan may feel more integrated into the neighbourhood than the insular Ferrier, but it is ironic that one scheme was bulldozed to make way for another with a 21-storey tower at double the density of homes of the old Ferrier Estate

(Scanlon et al. 2016). This shows the political reality of housing delivery practices, recasting the experiences of the past. Of course, the Urban Task Force and New Labour's Urban Renaissance policy, of which Kidbrooke Village is an example, has been critiqued as copying continental European models of apartment living and as a 'gentrifier's charter' that seeks to socially mix (in reality socially cleanse) council estate populations (see Lees 2003). It is interesting to note that a recent review of dense living in London by Blanc et al. (2020: 7) is ambiguous about the utility of the 'eyes on the street' notion of defensible space in these new dense developments: 'Attractive outdoor spaces with comfortable seating and convenient pedestrian routes were better used than hard surfaced, dead-end, heavily overlooked places'. This ambiguity is also evident in their interviews with residents of Strata Tower (which leaseholders on the adjacent and now demolished Heygate Estate in Elephant and Castle were recommended to move into, see Lees and White 2020):

> There isn't much interaction although there are a lot of people – so its secure in a way but also gives me my space … I like that its gentrifying, better people moving in is a good thing. (Blanc et al. 2020: 22)

Ironically, despite their promotion of tall buildings and increasing densities, New Labour policy makers also bought into ideas of defensible space. Indeed, in order to see the resilience of defensible space ideas under New Labour we only need to turn to DETR (later the Office of the Deputy Prime Minister [ODPM])[12] and the setting up of CABE. CABE was a small (for government) organisation with an unexpectedly high degree of reach and influence. It directly shaped the quality of English housing from 1999 through design review and hands-on enabling advice to Local Authorities, housing developers and government departments themselves. In its most active decade CABE was responsible for a prodigious output of advice, guidance documents and evidence to support urban design policy, especially for housing. While John Gummer, Secretary of State for the Environment in the Conservative Government of the 1990s, may have initiated the government's consideration of design as an area for public intervention, it was under New Labour that CABE set standards and codes of good practice for housing (through practical tools and awards such as Building for Life, the National Housing Audits of volume house builders' output, social housing and design code pilots, etc.). CABE's role and contribution to design governance is explored in detail in a publication explaining the positive impact of these various tools (see Carmona et al. 2017). This identifies the CABE experiment as a product of a period of rapid government expansion into new areas of public policy such as built-environment design, showing how state-led design governance was used to pursue related public policy goals. CABE's activity, however, was criticised as much as appreciated. It was accused, amongst other things, of being overly concerned with architectural aesthetics and portrayed as an urban centric, neoliberal, pro-development organisation. Its championing of increasing housing density was typically contentious, with the promotion of good place-making and better design seen as a cover to foist unpopular new homes on dis-affected communities. Yet a closer reading of documents such as *Better Places to Live By Design: A Companion Guide to PPG3* (DETR and CABE 2001) shows how CABE provided a balanced view of complex concepts, distilling them into accessible principles to address the gaps in understanding of planning policy for an audience of lay people as well as architecture, planning and construction pro-fessionals. In *The Councillor's Guide to Urban Design 2003*,[13] CABE defined defen-sible space for Local Authority councillors as:

> Public and semi-public space that is 'defensible' in the sense that it is surveyed, demarcated or maintained by somebody. Derived from [the] Oscar Newman 1973 study of the same name, and an important concept in securing public safety in urban areas, defensible space is also dependent upon the existence of escape routes and the level of anonymity which can be anticipated by the users of space.

Thus, by 2003 the Government's body responsible for promoting good housing design was highlighting the practical contradiction inherent in defen-sible space, asserting that for spaces to work successfully requires enclosure plus a degree of free egress and accessibility, rather than complete exclusion. A commitment in the Urban White Paper (2001) also led to the drafting of *Safer Places: The Planning System and Crime Prevention* (ODPM and Home Office 2004), which outlined seven attributes of safer places and the contri-bution made to feelings of community cohesion and safety. These are the, by now familiar, attributes:

- *Access and movement*: places with well-defined routes, spaces and entrances that provide for convenient movement without compromising security.
- *Structure*: places that are structured so that different uses do not cause conflict.
- *Surveillance*: places where all publicly accessible spaces are overlooked.
- *Ownership*: places that promote a sense of ownership, respect, territorial responsibility and community.
- *Physical protection*: places that include necessary, well-designed security features.
- *Activity*: places where the level of human activity is appropriate to the loca-tion and creates a reduced risk of crime and a sense of safety at all times.
- *Management and maintenance*: places that are designed with management and maintenance in mind, to discourage crime in the present and the future.

The main emphasis was on creating well-designed places from the outset, rather than an over-reliance on installing costly security measures as an afterthought.

Post-hoc security measures or an insensitive approach to housing estate layout was more likely to lead to places that are neither well liked nor respected. *Safer Places* highlighted both the generic and specific design principles for reducing the likelihood of crime in residential areas:

> the desire for connectivity should not compromise the ability of householders to exert ownership over private or communal 'defensible space' ... access to the rear of dwellings from public spaces, including alleys, should be avoided – a block layout, with gardens in the middle, is a good way of ensuring this. (ODPM and Home Office 2004)

The collaboration between the ODPM and the HO on *Safer Places* highlighted the different emphasis planning and crime science disciplines placed on the efficacy of housing design as a solution to the impacts of crime on things, places and people, communicating to different practitioner audiences. Foremost in practice-oriented crime academics specialising in housing is Rachel Armitage, who from her early research career evaluating SBD housing within West Yorkshire (2000) has researched and written extensively on community safety, anti-social and pro-social behaviour, and many related aspects of residential crime prevention: alley gating, CCTV, risk assessments of residential areas and burglars' decision-making. Following her doctorate, 'SBD An Investigation of its History, Development and Future Role in Crime Reduction' (Armitage 2004) she continued to provide SBD theoretical credibility and rigour, and to shape policies and practice that impact on housing design and planning to prevent crime (Armitage 2013). Influencing crime practitioners through the Designing Out Crime Association and being a frequent member of HO taskforces, Armitage is part of a growing academic movement to revive the relevance of CPTED by re-establishing its connection to other built-environment disciplines (Armitage and Ekblom 2019). Importantly, Armitage provided another transfer link between government departments when she was commissioned in 2010 by CABE and the HO to investigate the crime experience of contemporary housing schemes. Echoing aspects of Coleman's approach, this study mapped design features of newly built housing schemes against recorded crime data, identifying practical urban design and space management lessons.

Armitage et al.'s (2010) meta-review of post 2000 national and international crime reduction and housing-design policy found conflicting issues as well as areas of commonality between documents such as *Better Places to Live* and *Safer Places*. A foundational tenet found in both housing and crime policy was that the design of homes and surrounding spaces influenced crime levels. The basic principles behind defensible space (appropriate levels of surveillance, clearly defined ownership of spaces, comprehensible layouts and maximising the presence of individuals) were recommended in all of the documents reviewed. Two areas of policy disagreement covered issues that Coleman omitted from her DICE principles: parking and ease of through movement. Unlike earlier guidance, subsequent policy, e.g. *By Design* (DETR and CABE 2000), placed greater emphasis on the impact of cars in a domestic setting. There was agreement that certain car parking solutions attracted crime, but less concurrence on the appropriate location for parking.[14] There was significant variation in guidance documents around desirable levels of through movement and connectivity. This ranged from encouraging high levels of accessibility, emphasising the benefits of walking and cycling networks and integration to the surrounding areas (CABE 2009; DfT and DCLG 2007; ODPM and Home Office 2004) to limiting excessive permeability from layouts based on through routes (ACPO, 2010): 'The desire for connectivity should not compromise the ability of householders to exert ownership over private or communal "defensible space"' (DfT and DCLG 2007: 47). The difference here is one of emphasis rather than significance; all the guidance acknowledged that through movement was an important factor, requiring localised decisions on appropriate levels of permeability dependent on the context. What constitutes excessive permeability remains unquantified. This need to consider specific circumstances, using professional judgment to assess appropriate levels of movement, is an example of policy and guidance's reliance on *phronesic* knowledge (see Chapter 3 and Figure 3.7) to decide suitable actions on a case-by-case basis.

In another example of the transience of policy fashions (for evidence-based policy, as well as guidance derived from evidence), the current status of design and security guidance in British planning and housing policy continues to be uncertain. Two government reviews of housing and planning policy indicated that the 2010/2015 Coalition Government viewed design and planning guidance as an unnecessary hindrance on the housing market. Lord Taylor's (2012) *External Review of Government Planning Practice Guidance* recommended the deletion of a major proportion of the Department for Communities and Local Government's (DCLG) guidance, including planning statements, circulars and design guides. Taylor (2012) proposed that it was not the role of government to provide 'best practice' guidance. He based this on two arguments: that practice is continually evolving and so guidance dates quickly (some of the 200 deleted documents were over 50 years old) and that the various 'practitioner bodies' or industry organisations were better placed to provide more relevant and accessible web-based research and resources (Carmona 2013). Taylor (Lord) (2012) believed that the urban design advice contained in cancelled guidance such as *Safer Places*: (ODPM and Home Office 2004) or *Better Places to Live By Design* (DETR and CABE 2001) were now part and parcel of mainstream planning, urban design and architecture practice. In Taylor's view *phronesis* had incorporated *techne* (see again Chapter 3, Figure 3.7) to the extent that only the high-level designing for security principles needed to be retained in government guidance.

Lord Taylor triggered the *Housing Standards Review* of 2014, which was a radical streamlining of national housing standards and guidance that was seen to be over complicated and restrictive. The principles of housing standards such as the Code for Sustainable Homes, the Merton Rule and SBD were to be incorporated into a simplified compliance regime for Building Regulations (DCLG 2014a; Williams 2014). One outcome was a new section, *Approved Document Q: Security in Dwellings*, in the English Building Regulations in October 2015. This was limited to a subset of target hardening, the secure design of doors and windows for new schemes. These two reviews and the National Planning Policy Framework (NPPF) resulted in the most comprehensive restructuring (some say upheaval) of planning/urban/housing policy since the 1948 Town and Country Planning Act. Such fundamental changes were justified by the requirement to stimulate construction of greater numbers of *new* homes, but as the history of SBD has shown, the policy read-across from the expected quality of new housing to refurbishment is patchy. The latest edition of SBD in 2019 has been extended to cover all homes: not only newly built ones, but also refurbishment or regeneration projects. Intended to incorporate the legacy requirements of the Code for Sustainable Homes until a replacement was established, the introduction emphasises carbon and environmental savings from avoiding the replacement of poor-quality doors or stolen property. The document's three sections cover the design and layout of housing developments, a reiteration of Part Q (the 'police preferred specification' for physical security), and a final section on 'enhanced security features' required to receive SBD Gold Award compliance. This three-stage compliance indicates an attempt to retain the relevance of SBD (now awarded for achieving the minimum standards of the Building Regulations), at a point where there is an overall weakening of the explicit status of defensible space and physical design to reduce crime in the broader planning policy framework. While Homes England's priorities are now numbers of homes and accelerating delivery SBD is still included in their toolkit of methods, the 2019 *National Design Guide* refers to the National Counter Terrorism Security Office's *Crowded Places* guidance (2017) and *Protecting Crowded Places: Design and Technical Issues* (2014) but not SBD.

Some degree of weakening may be attributed to the erosion of resources needed to sustain a detailed evidence base. Following the governmental cuts in the 2010 'Austerity' Budget, the capacity for departments to provide the analytical or research evidence for policy formation was decimated. DCLG's budget was reduced more severely than other departments and, in 2011/2012, their budget for external research and consulting was 64% smaller than for 2008/2009. Between March 2009 and March 2011 DCLG's headcount of all research professionals was reduced by 30–40% and they lost 25–32% of their statisticians (Fenton 2012). The expectation was that under devolved localism powers, Local Authorities or even industry would step in to fill the void by gathering their own data and evidence. But reductions in budgets have affected industry and LA capacity as well. Use of SBD has only ever been voluntary, but its former inclusion in the *Homes and Community Agency's Housing Design Standards* (HCA 2007), or the GLA's *London*

Housing Design Guide (2010), exerted extensive pressure for its application. At its most popular, mandatory SBD use and a national ALO service had been suggested (most recently in September 2008) making the comparison to CPTED establishing itself as an international professional discipline (CABE 2008). However, resource intensity was used as an argument against this, as with only 13% of ALOs in 2008 working full-time, the police service would have been under-resourced to address the number of compulsory assessments (Armitage 2013). As a parallel consequence of the budget cuts, Police ALOs/CPDAs/DOCOs reduced in number from 347 across England and Wales in January 2009 to approximately 196 in 2012 (Armitage 2013: 208). So regardless of any need to maintain, or extend the existing evidence base, the numbers of experienced professionals able to interpret the guidance plummeted. As of 2019 the numbers of DOCOs seemed to have stabilised – the SBD website continues to quote a network of over 200 officers and staff liaising with police forces.[16] Regardless of government announcements on the end of austerity, cuts since 2010 have limited both central and local government investment in the activities to support SBD. The Ministry of Housing Communities and Local Government (MHCLG) was the worst hit of government departments with their Communities budget (covering main activities such as spending on social housing as well as administrative and research programmes) cut by more than 60% in real terms between 2010 and 2018, and local government spend on planning and development control falling by 41% in urban areas (Centre for Cities 2019[17]).

The era of DICE with its interdepartmental cooperation and collaborating crime reduction and housing research units now seems to have been a golden age for evidence-based policy. Occasional research studies have appeared (e.g. HO, Design Council, CABE (2014) *Creating Safe Places to Live by Design*[18]), but future government-led research on the topic from either the MHCLG (now DLUHC) or the HO looks unlikely. Barnes (2004a) described the struggles to incorporate research into government policy formation, but these difficulties have increased as departments no longer look to their internal research units to provide an evidence base, but to industry, commentators or compliant academics, in line with Young et al.'s (2002) elective affinity model of influence. In a situation where the research and policy interaction divide has become impassable, the result can only be ideology-driven policy.

Utopia on Trial Again

While it's politically unacceptable to publicly endorse ideas of social cleansing and gerrymandering, blaming the architecture and labelling it as failed is far less problematic in the political arena. Focusing on the architecture, Alice Coleman's 1985 book *Utopia on Trial: Vision and Reality on Planned Housing* illustrates how long this view has been held. Coleman links the ideology of social housing to the bygone era of post-1945 social democratic consensus politics by describing social housing as 'one of the great post-war visions', however, in the new neo-liberal environment, she saw the reality as one of 'squalor and social breakdown'... Architecture is blamed so that areas can be cleansed of social tenants (predominately Labour and non-registered voters[19]), who are seen as a drain on community resources and an irritant to those aiming to gentrify our urban cultural centres. (Davies 2017)

Now we have a prime minister who talks in very Victorian terms about postwar [sic] council housing estates. Once again, the expressed fear is that certain types of building harbour crime and rioters, so must be bulldozed. 'Ventilating the slums' is back. The language is emotive: 'bleak, brutal high-rise towers', 'dark alleyways', 'sink estates', 'pointless planning rules'. It seems that these estates 'designed in crime'. Most of the 2011 rioters came from post-war council estates, therefore the buildings are to blame. (Hugh Pearman in the *RIBA Journal*, 2016: 67)

Beyond the technocratic interventions of SBD and BS, we have shown how defensible space was strangely resilient in New Labour policy-making. But as New Labour exited the scene, the housing utopia was placed on trial again, as a consequence of the revived calls to dismantle high-rise council estates. This book highlights one way that the problems of England's large-scale council estates – especially those that were difficult to live in and to manage – have been on the political agenda since the late 1970s (see Figure 6.2). Defensible space ideas were

Figure 6.2   Aylesbury Estate tenants disagree with Alice Coleman's views. (Source: Simon Harding)

put into practice via many of the key estate-based regeneration programmes, such as the PEP, EAP and DICE, which we have discussed. Defensible space ideas were also subsequently embedded in more recent programmes, such as HATs, City Challenge, the SRB, LSVT and the Estate Renewal Challenge Fund. Defensible space also made its way into the Social Exclusion Unit's *National Strategy for Neighbourhood Renewal,* whose remit was to 'develop integrated and sustainable approaches to the problems of the worst housing estates' (SEU 1998: 3); including the NDC and the Sustainable Communities Plan; it was also evident in the debate around demolition of homes within Housing Market Renewal Areas. But this time social exclusion, rather than design, was the motivation, for how could 'flats and public squares be inherently bad, when the new planning policies courtesy of Lord Rogers and John Prescott insisted on them?' (Hatherley 2015). We have not had the time or resources to investigate these in any detail, having been limited by familiar constraints such as access, time demands and the costs of 'following' in policy mobilities research (see McMenzie et al. 2019). Nonetheless, the repeated framing of the problems and their solution as critiques of the form of council housing and spaces around them, demonstrates that defensible space has been accepted into the canon of housing and regeneration policy as an effective intervention.

Today what is striking is how more recent (Coalition and Conservative) government rhetoric has all but copied that of the Thatcher government, in the way the rhetoric of failed design has seeped through and re-emerged, especially in ex-Prime Minister David Cameron's sink-estate initiatives, promoting a 'new' environmental determinism. Perhaps this is not surprising given the reprised role of Michael Heseltine (who as we have seen in Chapter 3 met with Alice Coleman at KCL at the time he was setting up the Urban Renewal Unit) as an adviser on estate regeneration. Writing in the *Sunday Times* in 2016, then Prime Minister David Cameron preached:

> There's one issue that brings together many of these social problems – and for me, epitomises both the scale of the challenge we face and the nature of state failure over decades. It's our housing estates. Some of them, especially those built just after the war, are actually entrenching poverty in Britain – isolating and entrapping many of our families and communities. I remember campaigning in London as far back as the 1980s in bleak, high-rise buildings, where some voters lived behind padlocked and chained-up doors. In 2016, for too many places, not enough has changed.

He continued, as we quoted at the beginning of Chapter 1:

> step outside in the worst estates, and you're confronted by concrete slabs dropped from on high, brutal high-rise towers and dark alleyways that are a gift to criminals and drug dealers. The police often talk about the importance of designing out crime, but these estates actually designed it in. Decades of neglect have led to gangs, ghettos and anti-social behaviour. And poverty has become entrenched, because those who could afford to move have understandably done so.

His diatribe against modernist, post-war housing estates continued:

> The riots of 2011 didn't emerge from within terraced streets or low-rise apartment buildings. As spatial analysis of the riots has shown, the rioters came overwhelmingly from these post-war estates. Almost three-quarters of those convicted lived within them. That's not a coincidence.

The spatial analysis Cameron refers to is Space Syntax's mapping of the location of incidents and the addresses of convicted rioters involved in the London riots. Space Syntax's explanation of the findings is less dogmatic than the Prime Minister:

> Hillier's earlier work suggests that the proximity of riot activity to large post-war housing estates may not be the result of social housing in itself but the type of social housing: most post-war housing estates have been designed in such a way that they create over-complex, and as a result, under-used spaces'.[21]

These spaces become occupied by groups of unsupervised youths and it is this segregated activity that displaces everyday activity, permitting the rioting to start.

Like Thatcher before him, Cameron repeats Coleman's anti-modernist council estate mantra. Slater's (2018) tracing of the etymology and usage of 'sink estate' suggests parallels to the increasingly common usage of defensible space as a catchall term. Like defensible space, journalistic use of the phrase 'sink estate' emerged in the early 1970s growing with frequency in political debate during the 1980s. It took off post-1997 following Tony Blair's visit to the Aylesbury Estate in what Campkin (2013: 95) identifies as a watershed that he termed the 'sink-estate spectacle'. Slater records how Cameron's 2016 sink-estate speech created an even greater explosion in its usage, promulgating the term as a shorthand caricature for deprivation, worklessness, family breakdowns and welfare dependency. The phrase is used to foreground anti-social behaviour, tainting places as under an unfixable spiral of decline moving the narrative 'away from community, solidarity, shelter and home' (Slater 2018: 897). Framing council estates as failing (see Lees 2014, on the Aylesbury Estate as 'sin city') is being used as an ideological assault on social housing:

> The sink estate argument aids the polarization of positions on regeneration into two opposed stances. Those who say the primacy of decision making over the future of housing estates should sit in the hands of the people who live on them. Others who see housing estates as public assets, and that the duty of local authorities or housing associations to use these to ensure that they are used in the fairest and most efficient manner. House builders and housing associations are now inculcated in the reshaping of public assets. This shared duty extends to finding space for additional homes for a wider mix of tenures, which in circumstance of escalating austerity and unaffordability, now extends to include middle-income households. (Architect Andrew Beharrell in *Urban Design Group Journal*, 2017: 15)

Cameron's response to this challenge was to set up what became known as the Heseltine Panel, a 17-strong group[22] co-chaired by Lord Heseltine and the then Housing Minister, Brandon Lewis, who reported to the Prime Minister and Communities Secretary Greg Clark. Its remit was to develop a national estate regeneration strategy and work with up to 100 council estates to tackle deprivation and transform them into vibrant communities. Reviving the mantra that all council estates were poorly designed, Lord Heseltine said: 'Estate regeneration is key to transforming the lives of people living on poorly designed housing projects'.[23] The inclusion of individuals from the Prince's Trust (set up by Prince Charles) and Create Streets (Nicholas Boys Smith, who as we have discussed argues for traditional terraced housing and streets) underlined the anti-modern architectural credentials of the panel and their views on defensible space (see quote from Nicholas Boys Smith at the start of this chapter).

Heseltine of course had long been a supporter of the dismantling of high-rise estates. Back in 1984 he had been impressed by the remodelling of the Cantril Farm Estate in Knowsley, Merseyside, using defensible space ideas such as breaking the housing up into small neighbourly clusters and building low-level walls for privacy. Indeed, he saw it as 'potentially trail blazing' (Brindley et al. 1996: 115). But defensible space, the Heseltine Panel and DCLG's extensive plans for estate regeneration were overtaken by the drawn-out political distraction of Brexit and the more immediate and pressing consequences of the Grenfell Tower fire in June 2017. As Boys Smith said about the Heseltine panel:

> I'm not sure it achieved much ... It was a disappointing speech that Cameron used to launch it. That actually didn't help. There you go. (interview, 2018)

This book is not the place to repeat the detail of the Grenfell tragedy, or speculate on the conclusions of the Hackitt review of building regulations and fire safety (2018),[24] even though the subject of safe secure homes, stewardship and the impact of living at high density and concentrated occupation are at the heart of this book. Outcomes of the Grenfell Fire such as the hoped-for rebalancing of responsibilities between tenants and landlords, or a shift in policies, have yet to be seen, but already there is increased scrutiny of housing issues beyond professional institutions or circles. The cladding that caught fire on the Grenfell Tower was a facade erected to hide its modernist, concrete aesthetic, symbolic of poverty and council housing, in an otherwise gentrified neighbourhood in a hyper-gentrifying city.

At around the same time and within the context of a housing crisis it is unsurprising that the Mayor of London's (GLA 2014b: 59; see 2017 version too) Housing Strategy called for the 'vast development potential in London's existing affordable housing estates' to be unlocked through private redevelopment. To 'kick-start and accelerate' that process, the government launched a £150 million Estate Regeneration Programme of loans to private developers 'redeveloping existing estates' on 'a mixed tenure basis' so as 'to boost housing supply' and 'improve the quality of life for residents in some of the most run-down estates in London and nationwide' (DCLG 2014c). Launched on the same day and underpinning Cameron's declaration of action on sink estates was property consultant Savills' 2016 report to the Cabinet Office – *Completing London's Streets*. In this Savills (and co-authors, Create Streets) critiqued contemporary models of estate regeneration that largely replaced existing buildings with high-density blocks, comparing this to an alternative 'complete-streets' model echoing traditional mansion blocks and rows of terraces. Savills claimed this approach optimised space though efficient design, allowing equal or more homes to be built in a form that would be both cheaper to construct and to maintain. The report identified 8,500 hectares of land in London, in local authority ownership, accommodating 660,000 households. If these estates were rebuilt following the 'complete-streets' model, an additional 480,000 households could be housed on them. The number of additional homes realistically deliverable for the capital was subject to dispute, Savills suggesting anywhere between 54,000 and 360,000 more homes[25] against Centre for London's (2016) estimate of 80,000–160,000 at similar density levels. The 'complete-streets' model would also build shops, parks and other employment opportunities within a familiar street layout, making the development of a greater number of homes and commercial space more viable, overcoming the challenge of high land costs and resulting in long-term growth in value. Savills framed this as a more attractive proposition to landowners, landlords and residents, who all benefited from the value capture arising from a placemaking premium. Savills contended that long-term patient capital funding was essential, rather than short-term debt-reliant borrowing. They acknowledged that to achieve high-quality regeneration, the public sector required long-term income streams for both capital and revenue investment and called for a reconnection of rental funding models to landowner compensation.

DCLG extrapolated the arguments in this report from a theoretical exploration of land value capture in London to an argument for levels of funding required for a national programme of estate regeneration. The resultant Estate Regeneration Strategy (2016) set out an ambitious vision to redevelop estates, in order to transform neighbourhoods, improve quality of life and deliver thousands of additional homes, with the 2017 Housing White Paper setting out a commitment to 'support the regeneration of our cities, towns and villages'. However, while £140 million was originally pledged to unlock estate regeneration in 2016, only £32 million has been distributed, and none since early 2017. There was a shift in sentiment between the Heseltine Panel's (unpublished) interim report due to be launched in June 2016 and the tone of the National Estate Regeneration Strategy published six months later. The interim report led with the role of regeneration improving life chances, but this was overtaken by the Brexit Referendum vote and Cameron resigning as Prime Minister. Buried within the interim report is reference to a research study to understand better the public expenditure on targeted interventions and connecting estate regeneration with wider service reform. Like Coleman, Heseltine continued to be an advocate for the idea that regeneration schemes could both enhance people's life chances and potentially lead to efficiency savings for the public purse.

At the same time as the National Estate Regeneration Strategy was launched, the GLA opened consultation[26] on their draft *Homes for Londoners: Good Practice Guide to Estate Regeneration* (GLA 2016). It took 14 months of fraught debate and lobbying from housing developers, London boroughs, housing associations, umbrella groups like Just Space and other stakeholders, before this was published as *Better Homes for Local People: The Mayor's Good Practice Guide to Estate Regeneration* in February 2018 (GLA 2018). The shift in title is indicative of the extent that regeneration in London is now intended to be resident-led. Transparency of consultation, full rights to return, fair deals for leaseholders and demolition only if this doesn't result in any loss of social housing (by total area rather than unit numbers), are enshrined in the guidance, as is the mandatory requirement for a supportive residents' ballot if a scheme is to receive GLA funding. The reinstatement of ballots followed the collapse of the Haringey Development Vehicle, a public–private regeneration partnership opposed by left-wing Momentum councillors, and increasing political sensitivity around large-scale regeneration delivering inadequate levels of social housing. The reappearance of ballots in the GLA guidance prompted the *Financial Times*' cynical assessment that the Mayor was now more concerned with reselection prior to local elections and the 2020 (pushed back to 2021) Mayoral race.[27]

In echoing the rhetoric of earlier unsuccessful approaches, the 2016 Housing and Planning Act (The Housing White Paper) posed a diffuse and fragmented series of moves to resuscitate a failed housing market. It saw regeneration and the delicate process of revitalising the homes of often battered/beleaguered council estate communities as less of a bespoke mending of existing social fabric and their surroundings than a wholesale restructuring of a neighbourhood and its mix of inhabitants. Social housing policy over the five years from 2015 to 2020 has been overshadowed by general austerity measures and exclusionary welfare reforms such as the threatened removal of housing benefits for under-21s, which has increased the risk of homelessness for thousands of young people, or Pay to Stay, which would have excluded higher earning households from social housing (Lees and White 2020). The 2018 Social Housing Green Paper contained several steps towards abandoning or reversing the policies instigated by Cameron's short-lived government. These abortive attempts to undermine the funding of new homes for rent, residents' security and to ration access to dwellings at an affordable level included: a higher-value assets levy or the selling off of council housing to repay right-to-buy discounts to housing associations (a Policy Exchange think-tank idea, which was axed in August 2018); mandatory fixed-term tenancies for council tenants rather than lifetime tenancies (proposed in the 2016 Act but never brought into force); or initiatives like Starter Homes (quietly terminated in 2017 without a single starter home built). Chancellor George Osborne's 2015 Autumn Statement, which removed all grant funding for rented housing, was the pinnacle of homeownership-focused housing policy. This was reversed a year later in 2016 by opening up a programme allowing housing associations to bid for grants towards sub-market rented housing. The same year, 2016, also marked the emergent rebirth of council house building programmes, which was further encouraged in the Autumn of 2018 by the lifting of the cap on Local Authority borrowing limits to fund new build housing.

This book does not seek to offer a full account of the history of English council housing as that would require a much broader perspective than our focus here. What we do show is how the privatisation agenda and the dismantling of council housing was tied up with the appropriation of ideas on defensible space. As Cupers (2016: 185) says: 'Ideas matter—but only to the extent to which they are mediated, appropriated, and translated into action by individuals who might or might not be conscious of their actual effects'. Defensible space ideas 'propelled a specific course of privatization that, ironically and sadly, has left more people homeless than could ever have been imagined by the often well-meaning theorists of human territoriality' (Cupers 2016: 186). Incidentally, in France there has also been a parallel between the various discourses of French urban renewal policy and the urban design solutions proposed by defensible space theory (see Vallet 2006). In a similar way, French discourses on urban renewal do not address the question of security head-on, instead promoting more consensual concepts like 'urban quality' and 'residentialization', frequently incorporated into the redevelopment of public spaces (e.g. delimiting paths, green spaces and play areas; visibility and legibility) (see Gosselein, 2016).[28]

Defensible Space Is Subsumed into Policy Thinking

Slowly, defensible space as a distinct concept is being expunged from formal planning policy. For example, the latest 2016 *London Supplementary Planning Guidance for Housing* still refers to Crime Policy 7.3 from the 2016 London Plan, which in turn refers to SBD 2010. These references were not reinstated in the GLA 2021b *Good Quality Homes for all Londoners* or the newest version of the London Plan (2021). The revised NPPF (MHCLG 2021) downplays community safety as a component of the broader social aspirations of sustainable development. The planning system aims:

> to support strong, vibrant and healthy communities, by ensuring that a sufficient number and range of homes can be provided to meet the needs of present and future generations; and by fostering well-designed, beautiful and safe places, with accessible services and open spaces that reflect current and future needs and support communities' health, social and cultural wellbeing. (MHCLG 2021: 5, para.8b)

The short Section 8, titled 'Promoting Healthy and Safe Communities' similarly emphasises the role of inclusive safe places to promote social interaction over explicit defensible space. This retrenched scope shows the extent that actively designing-out anti-social behaviour has become subsumed into measures to minimise opportunities for terrorism or other serious security incidents.

Defensible space has similarly been erased from regeneration policy. The only discussion of safety or security in the *Better Homes for Local People* guidance (2018) is related to fire safety rather than safety against crime. Like the NPPF, the 2021 London Plan refers to the outcomes of defensible space as qualitative design aspects to be addressed in housing developments where the 'site layout, orientation and design of individual dwellings and where applicable common spaces should provide clear and convenient routes with a feeling of safety' (GLA 2021b: 130). The hierarchical status and interconnectedness of planning legislation, policy, directives and guidance (particularly in London) can be obscure and affected by non-coordinated cycles of review. Tracing the inclusion and elimination of defensible space principles through these gives an interrupted pathway in comparison to the repetition and absorption of its underlying factors within design guidance.

Defensible space, however, has not disappeared entirely from official guidance. *Building for a Healthy Life* (Homes England's 2020 iteration of the Building for Life housing design standard) continues to use the term and to refer to its constituent elements, recommending designing homes with:

> Defensible space and strong boundary treatments ... Front garden spaces that create opportunities for social interaction. Ground floor apartments with their own front doors and semi-private amenity spaces help to enliven the street whilst also reducing the amount of people using communal areas ... Consider providing terraces or balconies to ... increase natural surveillance. (Birkbeck and Kruczkowski 2020: 83)

A further illustration of the concept's resilience, if in somewhat obscured form, can be seen through the gaining momentum of Create Streets – a social enterprise and independent research institute that campaigns for better-designed places. Nicholas Boys Smith, founder of Create Streets, and as we have discussed one of Heseltine's panel, was cochair with Roger Scrutton of the Building Better, Building Beautiful Commission[30] whose 2020 report exerted significant impact on the direction of current English planning, as the institute's media coverage illustrates.[31]. Their first publication, *Create Streets not just Multi-storey Estates* 2013 co-authored by Boys Smith and Alex Morton[32] of conservative think-tank Policy Exchange, as we have indicated elsewhere, takes *Utopia on Trial*'s thesis that multi-storey housing creates a spiral of decline for its residents and that the vast majority of people want to live in houses in streets. This was followed in 2015 with Boys Smith undertaking an approving interview with Alice Coleman 'Alice in Wonderland',[33] which reiterates the story of DICE and a critique of post-1945 planning, which 'she believes, still exerts a vice-grip on the provision of new homes'. Many of Create Streets' proposals can be seen, in part, to be based on the ideas of defensible space, at a point when official planning guidance (and practice) is less explicit in its promotion of the principles.

Create Streets not just Multi-storey Estates told those who would listen that Britain was repeating the multi-storey errors of the past. Like Coleman's *Utopia on Trial* it attacks post-war, modernist (Le Corbusian), high-rise estates; indeed it relishes what it calls: 'The start of a counter-reaction against high density' (Boys Smith and Morton 2013: 19). Reiterating defensible space ideas, they state that:

> Even when you take account of social and economic status, tower blocks and estate based high-rise and multi-storey living are meaningfully correlated with social breakdown, crime and misery. Even in the best of conditions, they are hard to raise children in, tend to discourage close human relations and provide a myriad of hard to police, semi-private opportunities for crime often with multiple escape routes. (Boys Smith and Morton 2013: 29)

They point to Coleman's 'influential survey' that found 'very strong positive correlations between levels of litter, excrement, graffiti and vandalism and the presence of tower blocks' (Boys Smith and Morton 2013: 30–31). And without corroboration, claim that studies from other countries 'strongly substantiate these findings and show fairly clearly a positive correlation between high-rise living, crime and behaviour problems and a negative correlation between high-rise living and neighbourliness and pro-social behaviour' (Boys Smith and Morton 2013: 31). Of course, care is needed using these criticisms as automatic arguments for dismantling high-rise living. It is worth heeding council house historian John Boughton's (2018: 5) caution, when reflecting on the Grenfell Fire, to:

> reject the bandwagon criticism of tower blocks as such. Their story is a mixed one but tower blocks have provided decent homes to many and continue to form a vital component of our social housing stock. The attack on high-rises slides easily, and sometimes explicitly into an assault on the form and principle of social housing more generally that should be resisted.

This mixed story is also true of defensible space. While its trajectory is less obviously in the ascendant, it has been subsumed into policy and practice to a great extent and the effect of the concept is still widely discernible. After 20 years of successful embedding in practical advice there has been a dilution of explicit defensible space and CPTED principles visible in policy documents. This has been happening in conjunction with a resurgence in directive government advice on estate regeneration. Yet defensible space's underpinning mechanisms are still to be found incorporated as an aspect of positive sustainable development, holistic planning or appropriate design-led density. The definition and perceptions of defensible space have shifted through the grafting on of new ideas that have extended the range of the concept itself, incorporating the social concerns of second generation CPTED, or assimilating divergent concepts such as permeability in a form of bricolage. SBD extended the concept of defensible space to incorporate detailed security design of individual homes through target hardening and specification of components. The concept evolved through its application to new situations not initially envisaged (e.g. to high-density estates connected by walkways, to the public realm more broadly, or to deal with factors ignored by Coleman such as to parking). The translations and adaption of defensible space into new contexts can be read in two ways: first, adaptation to differing locational contexts, where having emerged in the high-density urban context, defensible space principles continue to be found in more suburban examples such as the new Essex Design Guide (2018[34, 35]). Secondly, as this chapter has shown, adapting to transient and variable political contexts. Finally, in tracing the diffusion and spread of the idea, this chapter has identified SBD and the BS as defensible space's most successful national policy mechanism, even when SDB is no longer mandatory. Despite its national-level erasure, defensible space remains highly visible in many local variants within Local Authority guidance and supplementary planning documents, reiterating it as a common-sense concept.

Post the global financial crash of 2008 the number of new housing units completed has been the prime driver for British housing policy. More recently (reinforced by Grenfell) came the rediscovery that quality of construction and ongoing maintenance is critical. The current reintroduction of placemaking, and housing quality and design, into the debate by recent housing and planning ministers has yet to demonstrate that this is more than a populist attempt to influence local communities and reduce nimbyism. At its worst this resurgence in the belief that good design can solve all nature of problems is a tentative revival of DICE-like environmental determinism.

Defensible space has been a highly resilient idea and has continued its foray into English housing policy and practice. This is due to its adaptability, its malleability and plasticity; it is quite simply, a middle-range theory (see Chapter 7). Several theoretical elements have contributed to our critical narrative on defensible space, and not only on how it moved – these are:

- seeing it as a weak theory/concept;
- as fluid and mutable;
- as an idea that has been in and out of favour (so moving though time as well as location/context); and
- as a spatial concept that is best described in social terms ('ownership', 'confidence', 'positive social interaction').

The fact that there is a gap between the theory of defensible space and the reality has been helpful to its resilience in that its ideologues are able to continue to promote it as a set of ideas, principles and values even though (perhaps even because) it has never become a full-blown policy. Like Schmidt and Thatcher (2013) we see the apparent weaknesses of defensible space as having an advantage. As the revised Aylesbury Estate public enquiry showed, Newman's and Coleman's 'defensible space' ideas continue to be held up as totemic with little direct reference to the ambiguity of their ideas. Indeed, it is the ambiguity of 'defensible space' and its continued relevance that, we feel, will stand the concept in good stead, even as it disappears from the detail of English planning documents.

The story of defensible space and how it interacts with council estate regeneration policy is a reminder of the failures of past interventionist housing policies and supports a growing sense that regeneration policy has to be locally grounded. The tensions between national regeneration strategies and London policy, which needs to deal with the capital's extreme housing, rental market and land scarcity/high value are clear. Yet concepts included in the London guidance are influencing similar planning and design approaches in Manchester, Birmingham, and other dense urban areas in the United Kingdom. Our criticism of English housing policy is that it is not sufficiently nuanced to respond to local housing markets, conditions or spatial vernaculars. Housing policy in general and design policy in particular needs to be locally contextualised. Hence our book's emphasis on practitioners' use of middle-range theories, such as defensible space, to deliver the locally specific policy interpretation that good housing needs. The concept's ambiguity of definition can be seen as a positive evolutionary trait (also visible in other interventions such as trickle-down and social cohesion – although these have very different genealogies), so applying it in a consistent narrative requires clarity over who's version of defensible is being described in this case, and how and why.

Notes

1   *The fall and rise of the council estate.* https://www.theguardian.com/society/2016/jul/13/aylesbury-estate-south-london-social-housing.
2   1974 *Horizon* documentary 'The Writing on the Wall' https://www.youtube.com/watch?v=9OMH7N_6nCE accessed 02/03/2019.
3   From ethnographic notes at revised Aylesbury Estate Public Inquiry, Loretta Lees, 2018.
4   From ethnographic notes at revised Aylesbury Estate Public Inquiry, Loretta Lees, 2018.
5   From ethnographic notes at revised Aylesbury Estate Public Inquiry, Loretta Lees, 2018.
6   From ethnographic notes at revised Aylesbury Estate Public Inquiry, Loretta Lees, 2018.
7   The Audit Commission (1986) report *Managing the Crisis in Council Housing* that incorporated the DICE principles was also published in the same year, and it was highly likely that two high-profile government supported committees would have had common participants.

8 As of 2015 it is no longer mandatory.

9 There have been six versions of SBD – the original in 1989, 2004, then New Homes in 2010, 2014, Homes v1 in 2016, Homes v2 in 2019.

10 This is hard to confirm empirically as SBD is no longer mandatory, and comprehensive records were never maintained by DCLG of numbers of new homes that use the SBD standard let alone for refurbishment projects, even though this question was raised in parliament (*Hansard,* 21 January 2013, col.76 W).

11 The architect and RIBA President, Ben Derbyshire has worked on many regeneration schemes including the Mozart Estate.

12 The DETR became the ODPM in 2001.

13 *The Councillor's Guide to Urban Design.* https://www.designcouncil.org.uk/sites/default/files/asset/document/councillors-guide-to-urban-design.pdf.

14 The *Manual for Streets* (DfT and DCLG 2007) criticises parking spaces directly in front of the property as it breaks up frontages and reduces surveillance, while SBD favours this type of in-curtilage parking wherever a secure garage cannot be provided (ACPO and Secured by Design 2010).

16 *The Success of Secured by Design.* https://www.securedbydesign.com/about-us/news/the-success-of-secured-by-design-police-scotland-s-stuart-ward-showcases-extraordinary-87-reduction-in-crime-in-secured-by-design-properties.

17 *A decade of austerity.* https://www.centreforcities.org/reader/cities-outlook-2019/a-decade-of-austerity.

18 Tellingly the research was undertaken in 2009–2010 but publication and dissemination was drawn out and delayed due to reduced resources and disinterest from the HO.

19 See Chou and Dancygier (2021) on displacement and electoral interests with respect to council estate demolition.

21 Space Syntax, *London, riots research.* https://spacesyntax.com/project/2011-london-riots.

22 The members were: Councillor Ravi Govindia, leader of Wandsworth Council; Nicholas Boys Smith, director of Create Streets; Andrew Boff, leader of the GLA Conservatives housing group; Elaine Bailey, chief executive, from Hyde Housing Association, which regenerated the Packington Estate in Islington; Paul Tennant, chief executive from Orbit Housing Association, which regenerated Erith Estate in Bexley; Tony Pidgley, chief executive of Berkeley Homes – a lead partner on various estate regenerations across London; Peter Vernon, chief executive, Grosvenor Britain & Ireland; Jane Duncan, president of RIBA; Ben Bolgar, senior director of the Princes Foundation; Dominic Grace, head of London Residential Development at estate agents Savills; Emma Cariaga from the British Land and Thames Valley Housing Association; David Budd, Mayor of Middlesbrough; Natalie Elphicke, chief executive of the Housing & Finance Institute; Graham Allen, MP for Nottingham North; Félicie Krikler, director at Assael Architecture.

23 *Heseltine launches panel of experts to kick-start estates regeneration.* https://www.gov.uk/government/news/heseltine-launches-panel-of-experts-to-kick-start-estates-regeneration.

24 This independent review of the building regulations and fire safety, led by Dame Judith Hackitt, took a wider remit than combustible cladding, considering the whole regulatory system for design, construction and management of high-rise residential buildings. It resulted in the report 'Building a Safer Future' (MHCLG 2018). This contained over 50 recommendations on the requirements for a more robust regulatory regime. In it Dame Hackitt called for a 'radical rethink of the whole system' and a

'universal shift in culture' to drive greater responsibility for building safety. While initially limited to high-rise residential towers, 2020 legislation ensured that aspects of the report are applied to a wider set of buildings.

25 The scale of additional housing achievable and the use the report was put to in the industry press was fiercely contested, leading Savills to add the following statement to their website: 'This report is necessarily theoretical and does not intend in any way to directly inform the practice of regenerating estates ... This absolutely should not be taken as a recommendation that estates can only be regenerated through demolition'. https://www.savills.co.uk/research_articles/229130/198087-0.

26 *Completing London's streets.* https://www.london.gov.uk/sites/default/files/draft-good-practice-guide-to-estate-regeneration-main-consultation-summary-report.pdf.

27 *London Mayor blocks public funds for unpopular regeneration projects,* 2 February 2018. https://www.ft.com/content/6a99cb2a-080a-11e8-9650-9c0ad2d7c5b5.

28 *Urban renewal and the 'Defensible Space' model.* https://www.metropolitiques.eu/Urban-renewal-and-the-defensible.html.

30 'Building Better, Building beautiful Commission'. https://www.gov.uk/government/groups/building-better-building-beautiful-commission. On the back of this commission Boys Smith was appointed to create a new government body Office for Place, to oversee the creation of design codes for neighbourhoods.

31 Create Streets, 'Coverage, 2021'. www.createstreets.com/front-page-2/media-centre/coverage.

32 Morton who has been policy lead at the right-wing think-tank Centre for Policy Studies and Policy Exchange, was Cameron's housing and planning adviser in the No. 10 Policy Unit, working on key policies in the 2015 manifesto.

33 Boys Smith and Wildblood (2014).

34 'Private Space'. https://www.essexdesignguide.co.uk/design-details/layout-details/private-space.

35 *The Essex Design Guide.* https://www.essexdesignguide.co.uk.

Chapter Seven
Defensible Space: A Common Sense, Middle-range Theory

Many estates throughout the UK underwent makeovers as a result of Coleman's studies … and the generally inconclusive results of these laborious redesigns in eliminating crime (poverty was not mentioned) helped discredit her ideas, as, more decisively, did the new popularity of high-density inner city environments with the middle classes – though her work does live on in the form of 'Secured by Design' assessments, where new developments are essentially vetted by the police to make sure they don't 'encourage' crime. (Owen Hatherley 2015)

We don't think of defensible space as this thing over there and something else over there. We just keep asking, either through enquiry or doing our own primary research, asking about the links between urban form and good outcomes for residents or neighbours. What's popular? What supports the development? What can you find in the data? Some things are very consistent. Some things are more variable. (interview with Nicholas Boys Smith, 2018)

Housing is far too often reduced to a homogeneous ideal, rather than an assemblage of numerous complex interwoven influences, experienced by its inhabitants in diverse ways. As Colin Ward (1976) reminds us, 'The important thing about housing is not what it *is* but what it *does* in people's lives' (in Coleman 1985a: 182, italics by Coleman). Similarly, defensible space embodies a series of contested propositions about social processes and behaviours, explaining only one partial aspect of what housing 'does in people's lives'. The social processes

Defensible Space on the Move: Mobilisation in English Housing Policy and Practice, First Edition. Loretta Lees and Elanor Warwick.
© 2022 Royal Geographical Society (with the Institute of British Geographers). Published 2022 by John Wiley & Sons Ltd.

followed in the empirical research in this book have been those of *doing* research, *making* policy, *practising* estate regeneration and *undertaking* policy/research evaluation. In attempting to understand these processes, the intent was to (re) normatise defensible space as an academic *and* a practical concept, by unpicking the debate that surrounded it, and the personalities, politics and views of the participants – delving deep to uncover their value systems and the obscured agendas behind their actions.

In spite of extensive re-examination, defensible space and the explanatory models for how it works are still only partially understood (Cozens 2005; Ekblom 2011; Poyner 1983). Multiple investigations have tested the concept to an extent that the constituent elements are taken as a given (Cozens et al. 2001a; Merry 1981; Moran and Dolphin 1986), strengthening its hold on the imagination of the various disciplines who have adopted it. In various guises, the principles are repeated in current design guidance (Essex Design Guide 2018; Peabody Housing Trust 2018: 39). The widespread acceptance of the principles has resulted in outcomes, some of which are very consistent, some more variable, reinforcing that there is still much more to learn. Nonetheless this resilience, despite the doubt and controversy surrounding it, supports arguments for the continuing salience and pertinence of the concept, echoing Bill Hillier's (1973), and others' (e.g. Ley 1974b; Reynald and Elffers 2009), belief that defensible space was, and remains, a very important (geographical) issue. Thus, the gap this research has tried to fill is not merely what is known about defensible space, but more the gap in understanding about how this knowledge was/can be used and applied. This means asking whether the DICE programme was a success or failure becomes more a question of policy analysis than scientific proof of theory. Yet, while not aiming to definitively refute or verify defensible space as an idea/theory, an investigation of theory was still important – in tracing the movement from idea/theory to evidence-based practice, and how theories can provide frameworks for action for the communities and individuals that were interviewed. As co-authors we have been unusually fortunate to be part of all four communities of practice under investigation in this book: we have worked as academics researching housing and urban regeneration programmes; as researchers uncovering facts and insights; as policy makers using this evidence for policy formation; and as designers following the direction and advice provided by policy guidance, using research findings to inform decision-making. This multiple positionality provided us a unique opportunity to reflect on contrasting modes of enquiry, from the academic geographer's (and scholar-activist) research lens that Loretta is familiar with, to the practitioner's value-laden culture of certainty and professional decision-making that Elanor encounters day-to-day within her professional operating environment within a housing association. As this book shows, inter- and cross-disciplinary discussions are very productive, but they are, as yet, few and far between in policy mobilities research; we would like to underscore calls for more and deeper interdisciplinary dialogue.

By using policy mobility ideas to frame our discussion we followed a disciplinary tradition of policy analysis/evaluation as much as a geographical one. Both bodies of knowledge have shortcomings, particularly in how the mobility of policies might differ from the movement of concepts into research or into practice. So, to consider how research was called on to provide 'evidence' for both policy formation and practical action, we focused on the use of evidence to support the five policy roles described by Hogwood and Gunn (1984) as: i) an aspiration or mission statement, ii) a set of proposals, iii) formal authorisation that legitimises proposals, iv) a funded programme, and v) a process or field of activity. That DICE – and by inference defensible space – fitted this multifaceted definition of policy, justified treating it as mutating pseudo-policy.[1] Uncovering this more nuanced idea of what constituted policy and the resultant more fluid relationship between policy and evidence altered the initial conceptual framework.

Figure 7.1 shows that we originally intended to look at the interaction of research and policy and their respective impacts on practice through the lens of policy, assuming (with the policy mobilities literature in mind) that all three positions shared a partial viewpoint on defensible space. But, examining the material gathered has yielded insights on the contrast between design as an emergent process against the fixed nature of policy-making. So although this common ground did exist, it was less well defined – each position having its own subtly different reading of the concept. This meant that the relationships were more implied and less direct than the notion of evidence-based policy might suggest, and that the interaction with practice (particularly when considering *phronesis*) was only partially permeable, resisting or accepting evidence and policy pressures depending on opportunity and circumstance.

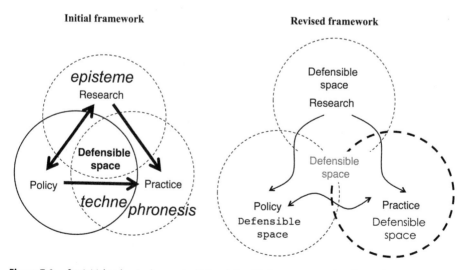

Figure 7.1   Our initial and revised conceptual interrelationship between research, policy and practice.

Mobile Concepts: Lessons for Academia, Policy Makers and Practitioners

The revised definitional framework (Figure 7.1) clarified the certainty, strength and directional nature of the tripartite relationship when answering the questions: *How did defensible space move? What was the role of transfer mechanisms? And, what were the varied disciplinary opinions on the value of evidence?*

In answering: *Who were the individuals and organisations who encouraged the idea of defensible space to spread?* the narrative arc of the story of defensible space in English housing policy and practice opens with the foundation myths created by Oscar Newman's research being taken up by Alice Coleman. This is followed by the struggle to obtain funding to test experimentally Coleman's redevelopment ideas, showing the problematic translation from academic research into policy-influencing research. As both research and policy formation, DICE reached a disappointing conclusion following the Price Waterhouse evaluation, which attempted to close down the avenues for defensible space within housing regeneration policy. So, the story also became one of 'how some ideas … persist despite premeditated or unintentional policy derailment' (Warwick 2015: 494). Defensible space (or Coleman's version of it) was described as an idea out of time, too anti-modernist and outmoded (Lipman and Harris 1988; interview with Paul Wiles, 2011); nevertheless, it resiliently re-emerged, exploiting not one, but multiple policy windows. Coleman's principles were temporarily more successful in other pseudo-policy formats such as the BS or SBD with elements still appearing in more recent policy documents[2] (GLA 2016).

Yet, as shown, the movement of the concept of defensible space into practice had in reality a longer, more complex and erratic trajectory. Starting with Newman's relatively simple recommendations for architects, the practical application mutated as it spread, via the DICE Design Guides for Housing Associations (Coleman and Cross 1995) or Sheena Wilson's advice for housing managers, evolving into the form recognisable in SBD, or the many policies of local planners (e.g. London Borough of Tower Hamlets 2009). However, more than written documents, the movement of the concept into practice is materially visible in the constructed examples, the growing numbers of (redeveloped) buildings by community architects and others applying and reapplying the approach (such as on the Lea View House estate in Hackney). Coleman's DICE schemes – particularly the Rogers Estate – played an important part here, leaving a built legacy that was visited by professionals (and indeed ourselves) and written about in trade journals. This gave the concept considerable traction, as architects are quick to pick up on talked about schemes, with the photos of a few familiar, well-designed, safe and secure remodelled and new homes[3] explicitly illustrating the design principles of housing associations involved in estate regeneration (Peabody Housing Trust 2018: 39, 53; see also

The Housing Design Handbook (Levitt and McCafferty 2018; co-authored by David Levitt of the community architecture practice Levitt Bernstein).

So, did the way in which defensible space was mobilised affect its impact? Unsurprisingly, considering the extensive discussion of impact in the policy mobilities and the idea-to-policy literatures, we found that the impact of defensible space was affected by the mobilisation mechanisms applied. Freedom for a concept to move is controlled by the nature of the transfer mechanism: be that an individual, a book or an event. The opinions of individual transfer agents (or policy entrepreneurs) often coalesce into 'boundary objects' (artefacts, documents, policy ideas or terms) whose historical emergence and take-up are occasionally easier to trace than the concepts themselves (Wenger 1998). Alice Coleman finding Oscar Newman's book in Canada reveals the book as a boundary object standing in for and transferring his opinions. More anonymous objects, such as the *National Audit Report on Council Housing* (which should have synthesised a range of stakeholders' viewpoints and hence provided a rounded view of what defensible space might be), relied on their organisational (or legislative) weight for their convincing influence. These mechanisms, particularly transfer agents, can act and influence uptake of a concept in positive or negative ways. Concepts can follow convoluted routes and may mutate or alter during translation. The receptiveness of practitioners is a critical aspect of successful embedding, as much as the adaptation or alignment of the policy to the new context. Alignment implies not just a single process, but that several favourable coexisting conditions are needed: persuasive mechanisms, unobstructed routes, timeliness and potential to fit.

The parallel attempts to transfer defensible space from the United States to England demonstrated multiple routes towards adjacent locations and audiences. At the governmental level, Oscar Newman's version moved (via his meetings with Sheena Wilson and others) from a US federal housing department (HUD) to a British security department, the HO. Coleman similarly acted as an international carrying agent[4] to the DoE, whose broader urban and housing concerns should have meant they were more receptive than the HO. Yet her timing was inopportune as the suitable DoE niche was already filled with an alternative response to the problem, the PEP. The practical and political conditions within the HO and the DoE acted to constrain, rather than stimulate, collaborative inter-departmental research. It took two transfer agents (Sheena Wilson and David Riley) moving between departments to fully mobilise information within these disparate organisations. Notwithstanding that Riley and the Price Waterhouse evaluation team acted as a resistive agent/barrier, it was the *practice* of interpersonal connections that produced a network of relations conducive to information flow, not just geographical co-location. This practice of connections and flows, with mechanisms following pre-activated networks, was repeated in the Tower Hamlets cases studied in Chapter 4. The conditions at the sites and within the teams working on the estates were more (Rogers) or less (Ranwell East) receptive to Coleman's version of defensible space. These examples suggest that mobile policies and concepts need more than an empty niche to move into and that the 'readiness' of the landing place and the adopters is as critical as the mechanism (cr. McCann 2013).

There is a high risk of failure from trying to mechanistically impose or translate an idea onto a misunderstood site. Kitchen and Schneider (2005: 279) summarise this insight:

> When looking at ideas from other places that may appear to be attractive, it is important to understand the context in which they have been applied and also whether or not robust evaluation has taken place. The failure to study both context and outcome can all too easily lead to ideas being imported which are imperfectly understood from the outset, and as a consequence can increase significantly the likelihood of failure when they are attempted elsewhere. In a world where knowledge in all its forms is not only constantly growing but is able to be moved around ever more rapidly, the risk of this is probably growing, especially where a 'quick fix' or 'something different' is being sought.

This identifies two further issues to consider: first, the implications of looking for 'something different' to transfer, and second, the speed of transfer. Healy (2010) has noted the importance of novelty of practice as a stimulus to transfer, over reinterpreting a familiar idea, but by the 1990s defensible space could not be considered a novel concept. An alternative explanation is that it was upheld by its strength as a 'common-sense idea':

> The language of broken windows, CPTED and defensible space has become so prevalent as to appear commonsensical ... this may be because it has at its roots certain taken-for-granted notions of space and property. (Blomley 2004: 633)

Common sense, ironically, refers to sound practical judgement that is independent of specialised knowledge; it is a kind of regular/normal/everyday 'native' intelligence on an issue. This of course sits well with the tenets of community architecture, where residents are seen to have good local knowledge about what design works or does not work.

The term policy *mobilities* also implies a swiftness of transfer. Concepts can move quickly when following the grain of advances. The societal context transformed rapidly from the 1970s when the idea of defensible space gained credence with planners and policy makers. Equally trends and 'fashions' in crime have evolved quickly, far faster than the policing response and certainly faster than the slow adaptation of the built environment. Housing and urban regeneration programmes are extremely long, drawn-out processes, hindering a speedy evaluation of the policies that initiated the changes. A persistent problem is the mismatch of timeframes for cause, intervention and impact. Decision-makers have notoriously short attention spans, looking for rapid policy effect rather than sustained long-term assessments of impact. Unlike housing initiatives such as Decent Homes, which began by modernising the interior of flats, Coleman was trying to work from the outside in. DICE improved the exterior of the estates relatively rapidly, transforming them in a highly visible way. Such obvious physical alterations may be politically desirable (e.g. the cladding on the Grenfell Tower), principally by benefiting the wider neighbourhood and 'tidying up' undesirable estates, however, as the Mozart Estate shows, building the community capacity to deal with social problems takes time, and slow steady phases of incremental change provide the time to ensure a sustained improvement.

Yet the failure of defensible space to evolve from a diverse and loosely defined concept to 'hard' theory was as much about how the concept was constructed, as how it was communicated. Understanding defensible space as a cluster of concepts explained the positive and negative consequences; the fluid toolbox of concepts provided resilience for the overall idea but the lack of any integrated model meant it became easy to unpick each fragment. The transfer mechanisms applied to each concept fragment altered that piece, helping it to align better into the unique context:

> Approaches need to be tailored to specific circumstances and to the people whose daily lives are framed by these circumstances, because the likelihood that there are standard formulae that can be universally applied with a guarantee of success is remote. The role of theoretical ideas and of experiences from elsewhere is to prove some starting points for this process, rather than to predetermine it. (Kitchen and Schneider 2005: 278)

This is a key insight that the theory or borrowed concept is the *beginning* of a process of local contextualisation (cr. Kennedy 2016).

Further constraints on impact arose from Coleman's role as a researcher attempting to influence evidence-based policy-making. Research often needs to be mediated in some way to gain traction and influence; for example, in this case, the National Audit Committee report. It is essential to take a realistic view of the capacity of research to influence policy directly. Successful policy influencing is all about timing, being in the right place, catching the eye of the powerful and having a zeitgeist-matching proposal. Influencing can take time and be achieved via unexpected routes. It is possible that Coleman had greater impact through her pithy, but timely, suggestion to Sir Kenneth Newman that the police could provide a crime-preventive assessment of proposed housing designs (Coleman and Brown 1985). Indeed, establishing the idea of advisors (proto-ALO's prefiguring SBD) proactively considering planning applications may have improved housing quality more significantly than all her other research and writing.

But what does this mobility of defensible space tell us about how these different communities of practice use evidence? An informative excerpt from the 1993 SNU report

Crime Prevention on Council Estates summarised the fundamentally different kinds of information needed to generate action by those researching crime prevention or for those implementing the advice in practice:

> Research into the relationship between design and crime is rooted in ideological assumptions and has often led to tentative conclusions with numerous qualifications attached. Unfortunately, when practitioners are considering design changes on estates, or in areas with high crime levels, the assumptions and qualifications are often forgotten. Publicity can turn specific research conclusions into an accepted common sense quite removed from a single study or drawing from a number of different, if not contradictory, theories. (SNU 1993: 157–158)[5]

What a researcher finds to be tentative contingent evidence, a policy maker may precipitately present as persuasive implementable facts, or a practitioner generalise into 'common-sense' accepted principles.

The diverse responses across disciplines highlight a challenge inherent to a multidisciplinary concept like defensible space: that it is exposed to unbalanced professional scrutiny focused on specific areas of expertise. Criminologists welcomed the introduction of novel (to them) spatial thinking but attacked Coleman's analysis of crime data, while architects dismissed the same spatial understanding as crude, revealing the distinctive ways that researchers, policy makers and practitioners interpret evidence. It is important not to consider these communities of practice as homogeneous or fixed, but to note that individuals can move between disciplines and connect networks. The main protagonists played several roles: Newman was an architect, planner and researcher; Wilson, a psychologist, criminologist and policy maker; Coleman a (physical then human) geographer with aspirations to influence both policy and practice; and Power a sociologist and housing activist, interested in urban problems, housing, crime and social exclusion. Each of these individuals' roles as transfer agents were constrained by their political, epistemological or disciplinary biases. There has been limited research on constraints on policy transfer/mobilities (it has mainly focused on demand, programmatic, context and application constraints, see Benson 2009, for a review) and next to none on how disciplinary bias can have a significant affect. But given that policy mobilities is about the power to produce particular forms of knowledge use (Cairney 2011), we need to ask disciplinary questions around their specific lens of belief and how each domain exercises power to compete with each other using disciplinary forms of knowledge.

The case studies of the Rogers, Ranwell East and Mozart Estates illustrate how the practitioner community is better than academia or policy circles at dealing with ambiguity in spite of the daily pressure for clear cut decision making.[6] This shows that *phronesis* is more than a superior form of practical knowledge, but is able to account for the choice between ends (the value rationality) and interests (power). Particularly illuminating here were the examples of ALO Mark Jones applying expert experience to interpret the codified SBD guidance, or Steve Stride[7] tailoring regeneration approaches to local circumstances (the local contextualisation of a national policy) by selecting (cherry picking) those approaches that might work, based on practical knowledge and local familiarity.

The case studies also show that selective defensible space principles might work in some places, but that outside these contextual issues design can only ever offer a partial answer. The ability of even a straightforward design to consistently deliver the solution promised is highly variable and subject to limitations; the management and maintenance of estates are now acknowledged to have an equal role to design in the process of sustained regeneration (see Baxter and Lees, 2009). The suspicion of design determinism demonstrates the widely held view that design alone is unable to provide definitive solutions, however good or careful the designer. Despite Coleman's faith in permanent design changes, just as design cannot be seen as the only cause of housing problems, it cannot provide the only solution. Coleman's belief in design primacy has been reassessed to a more realistic formulation – that some housing problems are intractable and can only be managed over time. Consequently, as there is no such thing as a permanent, generic solution to retrofitting housing, any proposed interventions need to evolve to match the shifting problems and situations, or available funding regimes.

The vocabulary of design problems and solutions is itself deeply problematic, implying the ability to identify and implement an optimal one-off, permanent answer, rather than a complex contingent response to the dynamic situation that exists. As Ravetz (1988: 162) suggests: 'the remedy of the problem estates is political and dynamic, rather than physical and mechanistic'. Roberts (1988: 123) ascribes the conditions that Coleman attributes to the breakdown of society and social malaise as merely failures of 'municipal housekeeping'. The inability to keep an estate clean, in good repair and rubbish or disturbance free points more to poor management than failed design or 'problem' tenants. There is a role for sensitive design interventions, despite studies showing that the presence of urban design characteristics such as integrated parking, a coherent street layout and well-structured wayfinding, are not good predictors of crime levels (Armitage et al. 2010). Pease and Gill (2011) conclude that as well as contributing to the overall quality of the housing development, design does influence security, but designing-out-crime cannot be reduced to good design.

If the case studies illustrate architects', planners' and housing professionals' tactical ability to function within a fluid local context, McCann and Ward (2013) criticisms of current geography policy mobilities writings argue that academia is less adept at the multi-perspectival, detailed immersion in localist policies that this requires. Contemporary policy mobilities literatures remain very abstract and theoretical. This methodological immaturity is hampered by the weak interest shown in how to apply these ideas in practice and the lack of experience within academia in how policy making actually works. Policy mobilities thinking provides a helpful model to track the movement of ideas and concepts, but it is still too weak an explanation to establish any predictable model. The literature tends to concentrate on specific issues, such as neoliberal globalisation, which limits their reapplication (see Jacobs and Lees 2013). Stone (2012) complains that to date policy mobilities studies have concentrated on the accessible, formal high-level state actors and 'hard' transfer mechanisms of policy statements and less on the slippery intangible processes, the informal meetings and interactions (Freeman 2012). This is changing, but slowly, reiterating Stone's (2012) belief that policy mobility merely repackages the ideas of policy transfer. In fact, policy mobility thinking shares many of the criticisms directed at evidence-based policy making. The discussion of evidence-based policy making similarly concentrates on what constitutes sound evidence; creating tools or models of 'what works' rather than understanding 'why this works' (or doesn't work) in this particular case. Older writings on policy mobilities usually looked at examples of successful transfer (e.g. Healey 2010; Peck and Theodore 2010a; Ward 2007). New work in policy mobilities is now looking at policy failures (see Baker and McCann 2018; Lovell 2017; Stein et al. 2015) suggesting that understanding failure gives a new focus on resistances and contradictions. In this book we have done just that, by showing activist/academic Anne Power to be a foil to Coleman's more purest approach in the dispute between community or design primacy, but we show this oppositional dichotomy to be contradictory. We have learnt from the occasionally unsuccessful transfer of defensible space, reinforcing the value of looking at 'poor practice' not just 'best practice'. Rather than ignoring the softer value-led conditions required for evidence to influence and persuade, policy making needs to be recognised as malleable and interpretative.

Many existing policy mobilities studies lack long-term temporal analyses. Here we have tracked the evolution and application of defensible space over a period of more than 40 years. This extended analysis has allowed us to show in some detail that processes of mobility and transfer are not simple linear progressions, but messy and contingent with multiple routes. Rather than following a single transfer route we have examined multiple and competing transfer mechanisms with respect to defensible space, looking at how they were at times mobilised and used in combination, and how politics on the ground influenced their movement. It was important to look at interlocking networks and to track how mobility is affected as it traverses domains (Larner and Laurie 2010), particularly as policies move as sub-policy fragments, or as pre-policy pieces (Peck and Theodore 2010a, 2010b; Jacobs and Lees 2013). Practitioners are practised at identifying, taking and reusing only those pieces that work. So, it is unsurprising that these were the pieces that moved. Such fragments got translated only in situ, reinforcing the importance of context, geography and the landing site.

Our contribution, in addition to reconstructing the complex political and personal story of defensible space and DICE, has also been in extending critique of policy mobilities to suggest new practical approaches derived from policy mobility/transfer mechanisms. This starts from the multiple mechanisms, routes and interpretations of defensible space that exist, noting how practitioners deal with these diverse versions of a concept within a particular context. These insights have traditional theoretical implications (questioning whether positivist scientific theoretical unity is achievable in practice), but also, we argue, useful practical implications in terms of triggering greater trust in practitioner experience based on a looser form of middle-range theory building. Policy mobilities scholars have not tended to offer policy recommendations for a whole host of reasons (see Soaita et al. 2021), but we can, as we do here, suggest new practical approaches that can speak back to theories and concepts.

Practitioners Develop a 'Middle-range Theory' about Defensible Space

The strength of middle-range theory is that it is not mindless empiricism and not abstract theory or theory about other theorists. Merton developed theory about how the world works. (Sampson 2010: 72)

Defensible space, we argue, is best described as a 'middle-range theory', a loose and ambiguous concept that is not yet a theory (although part of it is). Accepting defensible space as a middle-range theory accepts the ambiguity and complexity inherent in the idea, its messiness and critically its fluidity of practice. The idea of middle-range theory comes from sociologist Robert Merton:

Theories that lie between the minor but necessary working hypotheses that evolve in abundance during day-to-day research and the all-inclusive systematic efforts to develop a unified theory that will explain all the observed uniformities of social behaviour, social organization and social change. (Merton 1967: 39)

While middle-range theory is abstract and derives from empirical investigation, it is also generative and adaptive. As already stated, defensible space is emergent, taking an evolutionary fit to the particularities of a situation ensuring it is applicable to diverse real-world circumstances. It is these qualities that make it attractive (even without comprehending it as a mode of theorising) to practitioners. Nicholas Boys Smith of Create Streets acknowledged the appeal of theoretical certainty amongst the messy plurality of practical design and housing management considerations when interviewed in 2018:

How does space that's neither public or private, i.e. the corridors or stairwell, how does that work? What are the service charges that are required? What are the crime rates? Have they gone down? We're more interested in the more specific links between the discoverable statistical relationships between particular outcomes and particular design details rather than a corpus of theoretical work.

Stephen Marshall (2012) criticises urban design theorists' inability to scientifically test their ideas, identifying four ways that a theory can be pseudo-scientific. Coleman's fiercest critics accused her of the first two ways: being disingenuously fraudulent and/or deliberatively manipulative (Hillier 1986b; Lipman and Harris 1988) or being fundamentally implausible (Hillier 1986a). Even her less vociferous critics (Herbert 1986; Ravetz 1986) would agree that Coleman's positivism accords to Marshall's third situation:

> where the treatment appears to be scientific, where it may involve a method, approach or general mindset that 'appeals to observation and experiment' but which nevertheless does not come up to scientific standards. (Popper 1963: 4; cited in Marshall 2012: 259)

Most relevant to understanding how practitioners theorise defensible space is Marshall's fourth way of acting pseudo-scientifically 'as a sort of post-hoc justification for a form of practice that could proceed even without the theory' (Marshall 2012: 259). This post-hoc justification is visible in the DICE-like design changes applied to other estates (for example, the Tower Hamlets 'legacy' estates described in Chapter 4) and which were seen to work regardless of the theories associated with them. Marshall (2012: 246) continues that while urban designers do recognise and use evidence, the sector's inability to ensure that the 'scientific empirical evidential bases for its theories underpinning assumptions are correct, consistent and up-to-date' undermines the extent that evidence is valued. This failure is as true for housing design and management as a sector, as it is for urban design. The case studies show that Coleman's 'theory' of defensible space might explain some of the positive improvements that residents experienced, but equally so might other actions (better management practices, improved employment opportunities, changes in tenure or housing allocation). The architects and housing managers involved believed that the design interventions themselves might prove sensible, regardless of whether the effects claimed for them could be demonstrated (interviews with Sam McCarthy, Steve Stride and Nicholas Boys Smith, 2011; Ben Derbyshire, 2012; Elaine Bailey and David Gannicott, 2018). So, it should not be surprising that the practitioners interviewed repeatedly reported that whilst they valued evidence, they placed greater trust in experience (interviews with Mark Jones and Barry Stanford, 2011; Ben Derbyshire and Mark Hammill, 2012; Elaine Bailey and David Gannicott, 2018).

Whilst practitioners recognised the concept of defensible space, academics preferred to dismantle it. The SNU (1993: 157–158) quotation earlier in this chapter reinforces how practitioners create a recognised 'common sense' whole from separate (occasionally opposing) theories, with elements valued because of their communal acceptance – that they are 'held in common' (Wenger 1998). To practitioners, a fragment of the concept seemed valid if it chimed with their experiences, regardless of its source. Despite not having encountered defensible space during their studies, architects Mary McKeown and Peter Silver attributed their pre-knowledge of its principles to their practical housing experience.[8] So when they eventually read *Utopia on Trial*, their reaction was one of recognition and that the book articulated familiar aspects of their experiential practice within housing. As Robinson (2018: 234) says: 'the assumption that policy ideas even arrive is rather misplaced – at the very least we lose sight of the policies that were already there'. Pragmatically they tended to agree with the points of greatest overlap with their practical experience, even if this alignment of ideas was only partial. Both McKeown and Silver were less concerned with the political implications (or theoretical explanation) of Coleman's arguments. As a young community architect working on the Mozart Estate and looking for practical guidance, Alan Blyth was tolerantly open to the advice in *Utopia on Trial* (which he had 'lapped up' as a student, unconcerned by any controversy). He incorporated those of Coleman's principles that his clients and the residents responded to positively, alongside other approaches such as Bill Hillier's, fitting them into the SBD framework as required (interview with Alan Blyth, 2011). Those principles that didn't suit, he disregarded.

This openness, by practitioners, to selected aspects of Coleman's theories contrasts with how many of the academics interviewed remembered their introduction to her research (e.g. the interviews with Cooper and Woolley, 2013). By the time they encountered *Utopia on Trial* they were highly familiar with (or had contributed to) the earlier debate around Newman and determinism. Academics came to the work to critique or analyse; so rightly their focus was on the internal rigour of Coleman's ideas, the legitimacy of her use of data, as well as interpreting her overall propositions and political stance. Their interpretations were also more overtly shaped by how this matched their own professed position and disciplines. As academics they were less tolerant of Coleman's inconsistent thinking. Academic critique can be seen as more black-and-white, and less compromising, than the views of practitioners who can accept the spectrum of greys that ALO Mark Jones perceived. In addition, as Schön (1983: 37) recognised, academic novelty is valued more than utility with 'those who create new theory thought to be higher in status than those who apply it'. If theories are to be seen as boundary objects (transfer mechanisms) crossing the boundaries between researchers and communities of practitioners, Green and Schweber (2008) suggest that to be *useful* and to bridge the gap between these groups, theories need to have 'some degree of face validity, or a recognition by all of the potential utility of the theory'

(Green and Schweber 2008: 625). Middle-range theory building aims for just this practical utility: examining the local processes and problems and explaining the significant dynamics within a particular system, but with a relaxed acceptance of multiple and hybrid viewpoints (Green and Schweber 2008).

Key to this is accepting conflicting interpretations. Marshall (2012: 268) describes the simplification of urban design theories as a process of declining scientific clarity and influence: 'the original theories have "worn smooth" with time, losing some of their original nuance and purchase'. The opposite happened with defensible space, which became more baroque and convoluted as it was called upon to solve ever more complex problems. Clarity was lost due to this growing complexity resulting in a 'rats nest of intertwining hypotheses' (Rubenstein 1980: 6). As a conceptual framework, defensible space was based on several unresolved debates: territoriality as a positive or negative, open or closed layouts and resultant permeability, residential spaces as distinct from other types of public realm, and the degree to which environment influences behaviour. As a theory (if it can be considered one) it is inherently contradictory, with inconsistent effects dependent on context and inter-dependence of other features. A design may reduce crime levels but also reduce contact between neighbours; private gardens with stout fences may create defended buffer zones, but can also provide shelter for potential burglars. Yet our research shows that situated ambiguity can be a positive. Defensible space retained (and indeed still retains) its resilience because it is so fluid and ambiguous:

> It is useless to try to excise all ambiguity; it is more productive to look for social arrangements that put history and ambiguity to work. (Wenger 1998: 84)

Anyone encountering such a loosely defined concept can find some element they recognise, accept and apply, and can manipulate it to reflect their worldview. Thus, the idea becomes more complicated, incorporating later understandings and alternative interpretations. So, Peter Silver's view of defensible space was highly influenced by Bill Hillier's principles of movement through spaces. Silver sees spaces not as something that can be statically mapped (or definitely defined as semi-private, semi-public etc.), but as a fluid and relative condition, affected by permeability (interview with Peter Silver, 2013). Regardless of any theoretical difficulties this presents, Herbert (1982) and later Reynald and Elffers (2009) questioned whether the multiple versions of the concept are really a problem. Does it matter that the process of how defensible space operates is not well understood, or if it appears to work? Outside academic and policy circles, coexisting, conflicting interpretations of defensible space are accepted by practitioners, who are more prepared to follow the 'force of example' without much explanatory rigour. This can be seen as a common sense approach.

Of course, purity of theory in academic terms is pointless in practice. With such a confused and ambiguous concept as defensible space, beginning with the points of alignment and correspondence is perhaps more helpful than restating where they are in conflict. The conceptual basis of Newman's and Coleman's enclosed residential 'defensible enclaves', policed by occupants recognising and excluding strangers, is far from Hillier's shared residential spaces integrated within wider movement networks where strangers represent safety and opportunity for positive encounters (see also Jacobs 1961). Nonetheless, these contradict each other less than their authors might argue, and there are common characteristics where the detailed operational recommendations could potentially coexist (see Table 3.1 housing-design guidance devised from each of these theoretical positions showing where they overlap).

Newman, Coleman, Hillier, Wilson and even Power all saw physical design as a method of fostering urban civility; even if devising the design features to combine vitality, sustainability and security is challenging. A harder theoretical challenge is unpicking the deeply held assumptions beneath each of these versions. Hillier's 'safety in numbers' findings undermine assumptions that a large-scale high-density residential development is inherently more crime prone, or that community formation is an essential intermediary step in the creation of safe environments. He argued that 'simple human co-presence, coupled to such features as the presence of entrances opening on to a space, are enough to create the sense that space is civilised and safe' (Hillier and Sahbaz 2008: 27). Hillier and Sahbaz (2008) observed that the divergent principles that each side hold to be certain are all part of a larger, more complex picture. And as the complexity of a concept like defensible space is revealed, these principles need to be rethought. As such the ongoing arguments around defensible space provide another example that theoretical principles can only be the building blocks in constructing a stable concept, echoing here Forscher's (1963) metaphor for accumulating scientific knowledge as a wall constructed from the interlocking blocks of individual studies. In *Chaos in the Brickyard*, Forscher (1963) warned that concentrating on amassing these building blocks was not enough, paraphrasing Henri Poincaré: 'Science is built up of facts as a house is built of bricks; but an accumulation of facts is no more a science than a heap of bricks is a house' (Forscher 1963: 339). We would extend this metaphor further, with academic theory building being seen as a form of knitting, interlinking interdependent facts in an additive and highly controlled process, weaving in new strands into a useful theory. Whereas practitioner theory building is closer to a game of Jenga: constructing an edifice of all the currently accessible knowledge, then applying experience to dismantle this piece by piece, discarding those elements that are deemed unnecessary, inappropriate or unlikely to succeed in these circumstances. The skill here is knowing when to stop.

In the case study presented in this book the Mozart Estate is described as an example of *phronesis* winning out over *techne* and *episteme*. This talks to the role of experience in interpreting guidance. To paraphrase the ALO Mark Jones, practitioners are adept at taking what they need from a theory. They 'start small, start cheap, start local and consider the context' (interview with Mark Jones, 2011). But sensitivity to context can coexist with weakly evidenced decision-making based on heuristics or rules of thumb:

> Context dependence does not just mean a more complex form of determinism. It is an open ended, contingent relation between context and action and interpretation. The rules of a ritual are not the ritual, a grammar is not the language, the rules for chess are not chess, as traditions are not social behaviour. (Flyvbjerg 2001: 43)

In our research we found that failures often arose from unquestioningly following defensible space, DICE or SBD 'rules'; potentially undermining confidence that learning can be codified into best practice guidance applicable by expert or lay person alike (Wenger 1998). ALO Mark Jones pointedly distanced himself from sources of guidance believing his experience as an expert ALO held more power to convince than referring to others, regardless of their credibility:

> You have planners and policy and you have the ideas people, the gurus, the people whose books everyone reads. Realistically I think most practitioners of crime prevention and designing-out-crime would say they ignore those. They may read them initially when they're learning, but they don't go back and say 'Alice Coleman said'. And I don't say 'this is what Alice Coleman did' or anything to do with DICE or with anybody else when I'm giving practical advice. I give the advice that I know works. (interview with Mark Jones, 2011)

The same is true of policy. It is only in its interpretation and its application that its power is manifested. So, we need to carefully distinguish between policy making acts (the persuasive power needed in the moment of its writing and agreement) and the actions arising from the implementation of that policy (with their own convincing justifications) (see Freeman 2012). A strength of this research is our attempt to examine the gulf/space/tension between a statement of policy and its physical, embodied manifestation – the physical remnant that actually remains on a housing estate.

Throughout this research, the strongest transfer mechanism for defensible space been individuals, accentuating the power of personality. In earlier chapters we discussed Alice Coleman's role, more as an anti-guru, countering Peck and Theodore's (2010b) idea of persuasive gurus. Yet the strong and polarised memories of Coleman, a response to her 'marmite' personality, belie the significant impact and influence she had. Many interviewees commented on the force and abrasiveness of Coleman's personality, but Steve Stride identified the organisational challenge that she and her 'countercultural ideas'[9] represented to the DoE establishment:

There are these vested interests that will try and crush innovation. That's what happened with her. And Alice, they couldn't crush her because she was Alice Coleman and she had Thatcher backing her and she had £50 million and she had willing players like us. But directly they got the chance to snipe at it, they did. (interview with Steve Stride, 2011)

As discussed in Chapter 1 we were very mindful throughout this research of the ethical responsibility we had/have to Alice Coleman, as an elderly, female geographer. Such ethical responsibilities have seldom been discussed in policy mobilities research, but we feel we have presented a sensitive and balanced portrait of this important English housing (anti-)guru. We have certainly told HERstory and attended to her omission in the history of British (human) geography.

Unlike Morphet (2014: 40) who argues that 'an external "expert" may have more resonance than an internally generated idea', we have found ideas generated by 'outsiders' are frequently ignored or suppressed by the establishment either through fear of novelty or because of the threat they represent. Although aligning to the political right establishment then in power, Coleman can be characterised as a disruptive maverick and outsider, as much as a welcome innovator with privileged access to decision-makers. This unprecedented access (plus being foisted on the DoE as an interloper) undermined the established experts' organisational authority. She said:

My research never meant to offend but when one moves from one field to another (and land use is rich in fields) one sees things that conventional wisdom overlooks and some people cannot accept that their thinking might be wrong. (interview with Alice Coleman, 2008)

And Coleman was not alone in questioning the emerging evidence base on the interaction of housing design and behaviour. Bill Hillier, described as 'challenging conventional thinking since the late seventies' (Hillier 1998: 1), was also perceived as a pioneering, albeit disquieting influence. In 1987 Newman petulantly criticised Hillier's attacks on Coleman's and his own books as 'self-aggrandisement', accusing Hillier's dismantling of the concept of defensible space as a professionally irresponsible act. By encouraging housing professionals to disregard the concept he felt Hillier facilitated the construction of many substandard and crime-riddled estates over the subsequent 15 years (Heck 1987).[10] This is a sharp accusation, and blaming Hillier not only for the de-stabilising of a theory, but also subsequent poor decision-making, is harsh, while emphasising that the cut and thrust of academic debates have long-term practical consequences. In contrast, Anne Power's work, being located closer within (especially the New Labour) government and perceived to have a broader social-economic coverage and wider effect, has had a more sustained impact.

Considering the whole story, one striking observation is the extent to which the debate around defensible space and DICE stimulated the searching out of alternative interpretations. Kuhn (1970) argued that the continuing evolution of knowledge is reliant on periods of scientific revolution and paradigm shifts over-turning existing theories. Coleman herself articulated this paradigm shift:

> Modernism said, 'throw them together and they will form a community', but in practice they find anonymity. DICE said, 'give them freedom to control their own homes and then leave it to them'. What follows is a real increase in community spirit and a much greater and faster elimination of crime, etc., than I initially predicted. (interview with Alice Coleman, 2008)

Flyvbjerg (2001: 27) optimistically saw this theoretical disagreement as a prompt for alternative action, with divergent opinions pushing debate forwards faster than agreement, ensuring that the field develops in a knowledge driven manner. So, Coleman's success was to maintain interest in defensible space under immense critical scrutiny and attempts to discredit it. Her oft-described abrasive persis-tence promoting the concept was the 'grit in the oyster' that stimulates it to make the pearl. Tom Woolley, one of the conveners of the *Rehumanizing Housing* conference, reflected on her usefulness as a catalyst for forming consensus and agreement in a fragmented and argumentative sector. He gratefully recalls her acting as a cross-disciplinary rallying point for housing researchers, theorists and practitioners stimulating a united response against the rise of the New Right:

> In a way, Alice Coleman was helpful. If she hadn't existed, we would have had to invent her! (interview with Tom Woolley, 2013)

Other historical examples exist of thinkers who initiated or reinvigorated a para-digm changing debate, despite it later becoming clear that their views were almost certainly incorrect or perverse.[11] Coleman and DICE should be included in this category. In this sense it is not clear if defensible space is a success or a failure, we think it is probably both (and at different times and in different places). This underlines the utility of thinking about relational dyads, rather than oppositional dualisms (cr. Sayer 1991). The strongest cross-disciplinary consensus was a shared distrust of Coleman's design determinism – a charge she refuted (Jacobs and Lees 2013). Arguments over the relationship of design to behaviour (specifically here an-ti-social behaviour and crime) have been extensive and, as yet, inconclusive. Defen-sible space remains an ambiguous, slippery concept, with Coleman and DICE one heated tributary in the ongoing discussion. Whether or not her participation has long-term impact, Coleman's contribution to the debate was valuable, at least as a provocation to others: predominantly her belief that it provided a universal pan-acea to the many trying symptoms of 'social malaise'. That defensible space has moved into the mainstream, without definitive proof or consistent Government support, is due in no small part to Coleman's research. For all the uncertainty surrounding it, defensible space continues to be a powerful and influential way of salvaging 'problem estates', and making good failed housing designs and mistakes of the past. Coleman certainly thought so:

> It is good to know that the Dutch have now introduced effective reforms in the Bijlmermeer and made it safe[12] … I have also read how they moved obstreperous families out to a rather isolated location and built destruction-proof homes for them. Surprisingly (or perhaps not) these miscreants settled down and ceased to be a problem. This makes an important point. The DoE stressed the need to overcome socio-economic problems but no foolproof way has been found to redeem individuals in that way … Yet scientific design improvement really does work! (Alice Coleman interview, 2008)

Policy mobilities research and writing has tended to focus on policy success or failure (see Wells 2019, for a good review); our investigation of defensible space shows that it can be both and sets our study apart from more black-and-white discussions. More research needs to be done on the dialectic of success and failure and what it means in, and for, policy mobilities. Our discussion of a primary transfer agent and attendant transfer agents, plus the roles and work of many others, has also gone some way to disrupting (but not displacing in any spatial sense) the expert. Specifically, while we give Coleman status as an anti-guru, she was certainly not the only expert, as we have made clear – local communities and practitioners were/are the real experts on defensible space. Local interpretation is crucial, as the relationality of our policy sites, the United States and the United Kingdom, shows (Newman in New York City and Coleman, Wilson etc. in London). Oscar Newman saw British and European public housing design issues to be the same as those in the United States (as shown in his 1974 *Horizon* film), whilst Sheena Wilson and the DoE were strident that the British context was totally different. Overall, in terms of urban assemblage thinking in policy studies, our research demonstrates how well defensible space fits with McCann and Ward's (2013: 8) description:

> Policies … are not only local constructions; neither are they entirely extra-local impositions on a locality. Rather, policies and governance practices are gatherings, or relational assemblages of elements and resources—fixed and mobile pieces of expertise, regulation, institutional capacities, etc.—from close by and far away. They are assembled in particular ways and for particular interests and purposes … This concept [of assemblage] is helpful as a frame for policy studies because it emphasizes … that policies are not internally coherent, stable 'things' but must be understood as social processes.

Essentially, our argument is that defensible space is not internally coherent, it has been unstable across space and time, yet it may well be due to this that it has been, and likely will continue to be, resilient.

Research on policy mobilities is continuing to grow both in geography and beyond. This work has shown sustained interest in understanding the contextual inflections that influence policy mobility, the complex networks and relationalities involved, and the (re)making of policy as it travels between sites and people, if less so how it is put into practice. Soaita et al. (2021) state in their recent review of the application of policy movement ideas in housing research that there is no clear picture of how the key concepts underpinning policy mobilities research have been used to interpret housing policy developments. This book addresses that gap. But there remain questions about how we theorise and research policy mobilities, requiring deeper thinking about methodologies and research design. In this book we have tried to address both theory and methods, including questions of positionality and ethics. We have not focused on the familiar tropes of fast policy, neoliberal policy or policy gurus: as such, ours is a different read. Our research has shown that any stability in the policy assemblage around defensible space is provisional and limited, challenged in particular by the fluidity and mobility of the various actors. This opens up the question of what this impermanent malleability might mean for a policy mobilities field still very keen on the concept of assemblage.

This brings us to ask: *What might be the most productive next steps in policy mobilities research?* One step is investing more effort in trans-disciplinary research. Certainly, we have found that our trans-disciplinary approach, across geography, criminology, architecture and beyond, has been productive – and we would call for more work in this vein. Defensible space is a spatial concept, and these three fields each conceive of space and how it works in subtly distinct ways. These subtle differences matter and they risk being subsumed when considering defensible space – we may think we are talking about the same effect in the same way, but we aren't. Our trans-disciplinary research has shown us these multifaceted interpretations. It prompted us, and facilitated us seeing, the concept as multi-layered, with commonalities, in a way that allowed it to be appropriated and recombined in recognisable but in varied functioning ways. A second step is for academics working on policy mobilities to properly engage with policy makers and practitioners, collaborating together on research as we have done, thus bringing their different logics, temporal dimensions, communication practices, senses of rigour and relevance, interests and incentives together. Finally, given that 'it matters to the lives of many people what lessons from elsewhere are prioritised and what power relations they serve' (Soaita et al. 2021: 18), a welcome third step would be to investigate, in much more detail than we have been able to do in our research, the impacts of the mobilisation of policy on citizens, the non-elite or marginalised voices (or sometimes not), like those on the DICE-ed estates or Muriel Agnew on the Mozart.

Failing to Learn from Defensible Space

> Separation from the mainstream of transit and economic activity; the complexity and ambiguity of constituent spaces; the difficulty of navigation by outsiders; enclosure; covered entrances creating symbolic barriers or markers of ownership; the indirect relationship of street to home; and the complex and potentially illegible relationship between public and private spaces. (Severs 2010: 482)

Architect Dominic Severs' quote summarises the isolationist characteristics of estates alleged to have poor defensible space and connectivity. He was describing the Broadwater Farm estate in Tottenham, London, as an example of the 'characteristically high-rise, modernist and "non-street", the "no-go" estates of the modern era' (Severs 2010: 482). Broadwater Farm, completed in 1970 and cited in *Utopia on Trial*, became notorious as the site of riots in 1985 and again in 2011, and has undergone substantial waves of rebuilding and physical adaptation. Community consultation is underway for further estate improvement, including the demolition of remaining tower blocks, making, as with the Mozart Estate, the homes of around 4,000 people a site of continuous retrofit and reconstruction for many years to come, even opening the door to the displacement of long-standing tenants. The regeneration strategy prioritises safe and welcoming routes through the estate, active ground floors, overlooking improved open spaces, and to do this while rebuilding more densely, ultimately housing more occupants than the original Broadwater Farm,[13] in part echoing strategies proposed by Coleman. Yet despite the mobilisation of defensible space ideas in England, mass housing with Coleman's design 'faults' and 'poor defensible space' continues to be built (see Blanc et al. 2020, for residents' views on the positives and negatives of living in high-density schemes built during the last decade compared with more historical designs). Current schemes can be found in British cities that revive and exaggerate many of Newman's and Coleman's negative characteristics, perpetuating at least six design misperceptions about how the dimensions of defensible space interact and function as density increases. These illustrate a continued lack of understanding of how defensible space actually works and a failure to take on board the complex lessons that the critical history of defensible space teaches. Policy and guidance can only go so far to prevent design faults. The half-dozen shortcomings in design understanding with the potential for the negative impacts on inhabitants are:

1. *Design determinism obscuring other reasons for the failure of experimental, untested aspects of post-war housing design.* Typology isn't always to blame for failed housing. John Boughton's (2018) history of British council estates rejects a blanket criticism of the tower-block form, calling for greater attention to understanding the nuanced interaction of housing typology, height, density

and occupation levels. White and Serin (2021: 9) also argue that the height of a residential building matters less than the 'particularities of a building in its context', yet interestingly omit discussion of defensible space. The spatially inefficient and fragmented 'mixed-development' forms of post-war council estates combining towers and slab blocks were constructed in a period when London was depopulating, so did not consider optimum density levels (Boughton, 2018; Savills 2016). The mid-rise, mid-density terraces, typified by Sidney Cook's mid-century schemes in Camden, took a more sensitive and collective view on dwelling density and delivering shared spaces (see Swenarton 2017). Ultimately this idea of fluid, shared space was not always successful, as Bill Hillier and Space Syntax's critique of the Marquess Estate in Islington showed; censuring its tight, blind alleys and maisonette entrances raised up a flight of tight steps above blank walls. Also, the positive aspects of these exemplar designs were hard to replicate in the hands of less-experienced designers. The popular versions of these 1960/1970s experimental terraces were often successful because of their location and occupants – e.g. The Brunswick Centre in Bloomsbury and Branch Hill in Hampstead[14] – or their manageable human scale (the Alexandra Road Estate, Camden, while popular, is almost too large, with the Estate's streets struggling to relate to its context). This popularity based on location and scale is echoed in the reinvention and revitalised acceptance of high-rise towers in London (see Baxter and Lees 2009), such as the Trellick Tower and Silchester Estate in Westminster, or Holly Street and the Nightingale Estate in Hackney; suggesting their earlier problems to be primarily one of the experimental nature of the social interaction and urban form, rather than the (admittedly significant) failure of construction methods or investment in maintenance. Noticeably, it isn't only in London that new and refurbished high-rises are being promoted. Following the completion of the Beetham Tower in 2006, Manchester's skyline is being transformed by a burgeoning number of high-rise residential towers, with at least 13 residential towers of height 100 m plus either under construction or planned for the city centre as of Spring 2018 (Savills 2018). The Mayor of Greater Manchester, Andy Burnham, argues that towers and higher densities in the most accessible brownfield locations optimises land required for his promised 50,000 affordable new homes. While Manchester's forthcoming housing strategy will rely as much on suburban terraced housing as on these highly visible, new, city-centre apartment developments,[15] land availability drives density and height; pressures to preserve the green belts around Manchester and other English cities may make more homes in tall towers increasingly attractive.[16]

2. *Both post-war mid-density and modern super-density estates suffer from Newman's confused territoriality, accentuating disconnects to local milieu.* These often have forms and layouts that fail to follow the positive lessons of public street patterns, lack distinct front doors and entrances, with cars either ignored by the

design or poorly integrated, leaving the resultant schemes standing alone and not part of a permeable and well-connected streetscape. As many commentators, with as politically diverse views as New Labour's Urban Task Force and Create Streets, remind us, more traditional street patterns can deliver similar densities, with defined and manageable individual space. There is a call to return to familiar, traditional layouts, after the experiments in novel complex forms and spatial interactions that were being tested after the war. Not only were the outcomes of overly complex layouts initially less successful than hoped for, there was the additional failure of these forms to adapt to the changing demands put upon them, as the Mozart Estate case study shows. Older, simpler, estates have tended to adapt better: for example, the brick-built inter-war London County Council walk-up blocks. The benefits from designing for increased connectivity and permeability are being heeded, but clarity of hierarchy for the public realm and territory gets harder as density increases.

3. *Expecting that clever design of high densities/super densities can mitigate all the negative impacts of increasing levels of occupancy.* A basic lesson from the past 70 years of housing form is to be sceptical of experimental typologies that intensify one variable (density, height, social interconnection) without acknowledging the impact of interrelated influences. Yet this lesson is being overlooked. Current London planning policy and design vernacular is leading to the creation of blocky, overscale, super-dense, convoluted layouts with minimal public space – as equally challenging to manage as the complex interlocking slabs linked by walkways of the 1970s. The lack of critical thinking about how these newly built forms will be used suggests we are over optimistic in believing that careful design and defensible space will mitigate some of the worst problems. The small floor plates of tall point blocks may limit the numbers of close neighbours,[17] but the sheer scale and numbers of occupants, the pressure these increased numbers of residents place on neighbourhoods – both the physical infrastructure of amenities and green spaces and the social infrastructure of services and employment – should not be left unheeded. More neighbours means increased opportunities for anti-social behaviour and crime as well as increased opportunities for positive encounter.

4. *High levels of occupancy increase pressure on territoriality and reliance on mechanical, not natural, surveillance.* With the increasing numbers of microflats and co-living schemes, and increasing numbers of occupants in a single block, the application of defensible space no longer stops outside the front door of a flat, meaning Newman's criteria of semi-public and semi-private space cannot be limited to external spaces, entrances or communal corridors. The Collective in Old Oak Common, West London, is, at 550 rooms, the largest co-living scheme in Europe.[18] Individual private space is limited to a 11 m^2 en-suite bedroom with shared living spaces and amenities (co-working areas, bar, spa, gym, games room, library cinema, gardens, kitchens and dining rooms) accessible to all. Like many new housing blocks there is a communal

roof terrace and a 'secret garden', potentially used by all cohabitees, watched by all and under constant CCTV surveillance. The arguments put forward for developments like this are housing need and affordability, justifying a dispensation from UK government space standards. If these 'studio flats' had been built to minimum space standards (37 m² for a flat in the 2020 London Plan) around 100 fewer rooms would have been included in the building's envelope. The public communal facilities are extensive, but so are the long grey painted internal corridors. While some residents enjoy the hotel-like feel, a frequent criticism is that the 10-storey modern building feels less like a home and more like the overgrown student accommodation that The Collective was originally meant to be.[19] One visitor found the 'public spaces were eerily empty, with no-one in the shared kitchens, spa, library or corridors … Noticing the cameras around the place and signs saying "Smile you're on camera"'.[20] Amongst this large population, anonymity or respectfully detached reserve is just as likely as informal encounter: the scale of the scheme privileging mechanical over natural surveillance, curbing communal ownership of shared spaces.

5. *Limiting access to the 'public realm' creates segregated not communal spaces.* Pseudo-public spaces around housing are becoming frequent, with the segregation of external open spaces, demonstrating a negative version of surveillance and overlooking, where occupants can see into spaces but are excluded from using them.[21] The complaints against Guinness Housing Trust and the developers, Henley Homes, over children's play spaces at Baylis Old School in Lambeth shows this. Children living in the social housing part of the site were barred access to the play area when gates, included in the planning application (which stated the 'common areas are there for the use of all the residents'), were replaced by an impassable hedge. Following media condemnation, and Housing Secretary James Brokenshire's intervention, access to the play area was renegotiated.[22] Studies from the Joseph Rowntree Foundation and Chartered Institute of Housing, as early as 2006, argued that high-quality, mixed-tenure developments can be managed successfully and that developers should be willing to accommodate mix (Bailey and Manzi 2008; Rowlands et al. 2006). They recognised the risk that mixed-tenure estates might be difficult to sell or that property values are affected but that this could be eliminated by maximising the quality of other aspects of the development, including design, location and build quality. The report *Tenure Integration in Housing Developments* (NHBC & HCA 2015) found that while financing mixed-tenure developments could be challenging, property prices should not be reduced by integrating social housing. Planning policies encouraging mixed communities through the tenure-blind, pepper-potting of units and discouraging obviously downgraded entrances are now the expected norm in English cities. Nonetheless poor practices – for example, different external finishes, poor doors, or the relegation of social housing to less desirable parts of sites – still occur. In the 2018 revised Aylesbury Public Inquiry, it was noted with disgust by the objectors that the

supposedly tenure-blind, high-rise, new-build developments replacing the to-be-demolished Aylesbury Estate, specified a gold colour finish to the private block, but not for the social-rent block; and that, although there were no poor doors as such, the adjacent private and social rent buildings would have separate doors – segregating, not facilitating social mixing.[23] Drivers for such segregation are often a desire to make the social housing affordable to tenants as well as private housing more attractive to a certain clientele, recognising that different types of residents may have different priorities or needs and that it is reasonable to have variable charging regimes. Some aspects of defensible space do increase capital costs; numbers of lifts, controlled entry and CCTV also impact on ongoing maintenance costs. Additional facilities have to be funded, increasing service and management charges further.

6. *Segregation and distinctiveness as a desired attribute; the image of modernist tower blocks fuelling gentrification.* Of all Newman's dimensions of defensible space, image is the one seeing the most extreme recent reversal. The rekindled fashionability of brutalist concrete means the modernist typologies, the towers, walkways and communal spaces are now also deemed acceptable under a kind of image-led gentrification. Where once construction and material differentiated estates as 'other' now this distinctiveness signals exclusivity and desirability. Balfron Tower in Tower Hamlets (see Figure 7.2)

Figure 7.2   Balfron Tower, Tower Hamlets.

is a telling example of a housing form intended to foster communal living being subverted for economic and political reasons, with its iconic design status misappropriated to justify the shift. Completed in 1968 and listed in 1996, Erno Goldfinger's Brutalist Balfron Tower is a 27-storey slab with a detached service tower, comparable in appearance to his similar Grade II listed Trellick Tower in West London. The service tower contained communal laundries and 'hobby rooms'. Concrete bridges link across at every third floor and flats are combined with two bands of maisonettes, at ground level and on the 15th floor. Goldfinger saw these less as 'streets in the sky' than communal pavements, preferable to the limited opportunities for encounter provided on each floor of a point block. Coleman would have approved of the clustered design of the front doors, with small numbers of close neighbours grouped together to encourage interaction. Ownership of the tower and the surrounding Brownfield Estate transferred to local housing association Poplar HARCA in 2006. The basic concrete structure was in good condition and major remodelling and investment was planned to enable existing social residents to remain. But this was re-evaluated after the global financial crisis of 2008 and by 2010 Balfron Tower was beginning to be emptied and temporarily occupied by artists from Bow Arts as property guardians. This 'art-washing' of the decanting process continued until the decision was announced in 2015 that no social housing would be retained and all former social residents would be rehoused elsewhere. As property values across the east of the capital have risen, the vogue for Brutalist concrete architecture has returned (see also Davidson and Lees 2010 on Aragon Tower, now the Z Apartments, in Deptford). In May 2018 Poplar HARCA started selling off all the 152 refurbished flats (at prices ranging between £365,000 to £800,000, plus annual service charges in the region of £4k[24]) to maximise funding to build 130 new homes for social rent. In doing so Poplar HARCA has reinvented an exclusionary mono-culture, which is socially homogeneous, not mixed:

It's a symbol of the new world of social housing – new build financed by the construction of homes for sale and the mantra that mono-tenure (i.e. working-class, social rented) estates need to be 'improved' by an injection of middle-class affluence and aspiration. (Boughton 2017)

What is clear is that Bulos and Walker's (1987: 20) prediction for the future of (council) high-rises is already upon us: 'they will lose any remaining notion of being "streets in the sky" and will instead be viewed as the smart and convenient way of catering for the requirements of busy people who can afford to pay the heavy charges of "porterage" and design'. The local neighbourhood milieu around Balfron Tower is gentrifying equally rapidly. In the surrounding modestly designed Brownfield Estate, new-build houses with gardens for older residents of the tower are appearing, squeezed between two lower rise versions of Goldfinger's concrete blocks – Carradale and Glenkerry Houses. As part of the current refurbishment of the Estate, paved semi-public 'areas' have been built at the front of homes in a move that Coleman would have recommended. The children's play areas around Balfron Tower may have been redesigned, but it is no guarantee that children from the Brownfield Estate will benefit from the facilities.

In all of these cases it is the residents who have to live with/within the design failures. When practitioners disregard the warnings of defensible space, forgetting what housing 'does in people's lives', then the consequences fall on occupying families and households. Repeating the so-called design mistakes of the past inherent in various strands of defensible space (whether extreme occupancy levels, sky-high living, omitting natural surveillance, separation/segregating access to 'communal' spaces, or even using revisionist architectural fashions to drive mono-tenure gentrification) risks financial and personal costs to rectify them. The twentieth century 'design defects' Newman and Coleman identified have evolved to reappear in new housing in the twenty-first century. While this demonstrates yet another way that the concept of defensible space has persisted and adapted, reinforcing its resilience as a theoretical concept, it is *how* it continues to shape current practices of building new homes or regenerating estates that matters. Planning, design or crime reduction policy has been unable to ensure that what is built takes notice of the lengthy and extensive scrutiny given to the idea and its effects. As academics, policy makers and practitioners, we all need to reflect on the practical outcomes of our theoretical intentions. However much defensible space may appear simple and common sense, dictated by pragmatic rules about physical variables as apparently straightforward as height or access or adjacency, these interact at a social and behavioural level to a degree of complexity that we are still only beginning to understand. As these six headline examples of current practice demonstrate: get one aspect of defensible space wrong and the negative consequences may be rectifiable, ignore them all and a spiral of decline is inevitable.

Addendum: An Emerging Future for Defensible Space

In the final phase of finishing this book a global pandemic hit and discussions of public and private space rebounded as lockdowns and social distancing measures were imposed. The World Health Organization even listed Oscar Newman's 1978 *Defensible Space: Crime Prevention through Urban Design* under Covid-19 resources. And in thinking about what (US) cities might look like after Covid-19, William Fulton, Director of the Kinder Institute for Urban Research at Rice University, said, reinforcing the significance of careful design adjustments,

During the suburban era, we tried to solve most of our land-use planning problems by putting more space between people and buildings. But in cities, that's not possible. Instead of spacing our way out of problems, we have to design our way out. The threat of infectious disease is likely to ramp up urban design as a solution – perhaps, for example, by creating more separation in public spaces like restaurants and parks.

So, it's reasonable to assume we will begin to see small changes in urban design that separate people a little more and help make it easier to protect them.

Maybe the best analogy is the 'defensible space' movement started by urban planner Oscar Newman during the nadir of urban life in the 1960s. At a time when parks and public housing projects were crime ridden, Newman's view was that all space belonged to somebody. By giving residents a sense of ownership of these spaces, he contended, gangs wouldn't take them over. It was a revolutionary theory in urban design that changed everything – showing that, instead of fleeing cities, standing firm and redesigning them can be a better solution.[25]

The global design and architectural firm Gensler has also begun to argue for different approaches to open space post-Covid-19:

Seeing the way that people have flooded parks, greenways, and bike paths in recent weeks, overcrowding them and at times forcing authorities to shut them down, one wonders whether the current size, amount, proximity, and interconnectedness of open space is addressing our needs for today, let alone tomorrow. COVID-19 has caused us to develop a more nuanced appreciation of open space's social and therapeutic potential, such as access to better air quality and more physical activity, among other benefits ... Safety and security now takes on an entirely different meaning, one that goes well beyond Oscar Newman's Defensible Space theory. Urban planning must now push forward the idea that safety in the public realm is an investment in public health, thereby requiring us to think differently about open space in the face of this crisis. Cities and their local governments must embrace and manage security in our public realms to address this crisis and future pandemics.[26]

As we work out the practicalities of moving around cities and towns, during, and after, the pandemic, rapidly retrofitted street space, emergency cycle lanes, widened pavements, deserted public transport and more walkable mid-use neighbourhoods with well-integrated housing seem hugely desirable. The practices of negotiating space on pavements to take exercise or queue for essential groceries has forced us all to rethink ideas of territoriality and safe encounter. Walking through uncannily quiet residential streets, the aspiration to reclaim space from cars seems possible, and passing by households sitting on front steps – a reminder of Jane Jacob's North End stoops[27] – green communal front gardens in place of parked cars where neighbours can gather and children play are easy to imagine.

Turning off the terraced streets of homes with their own front doors, onto estates or high-density blocks, past fenced-off playgrounds, we realise how valuable the left over 'wasted spaces' have become. Courtyards have become spaces to sit out amongst stranded and unused cars. Doors and windows are left open, renegotiating a sense of privacy and ownership, and appropriating what would have been thought of as public realm for social [-ly distanced] activities.[28] A wandering stranger is waved through, acknowledged with a nod even if their face mask obscures a smile of greeting. As our personal territories have been curtailed during covid the subtleties of ownership of intermediate spaces are blurred (see Warwick and Lees, in review).

Where space within our homes feels constrained (even claustrophobic for some) by all the additional activities lockdown has required – space to work, schooling at home, learning how to come together as a household (again), or needing space apart – it puts even greater pressure on semi-private spaces. Access decks have become not streets in the sky but living rooms beyond the front door. For individuals living in isolation the semi-public spaces where you can confidently stand on a balcony to clap, or wave at a passer-by, have been a lifeline for mental wellbeing and social connection. Everyone needs some outside private space, but we have learnt we need the transitional spaces just as much.

The number and proximity of our neighbours, who has right of access to spaces, who we know and trust, is undermining architects' familiar rules of thumb:

> As a general rule, not more than twenty flats sharing a secured entrance, with no more than eight flats per floor, works satisfactorily but the increases in height and density of flats reliant on lifts makes these targets unreachable. (Levitt and McCafferty 2018)

Hillier's (1988) advice to avoiding clustering too many entrances in too small a space seems prescient. How many people we share an entrance or a lift with, the dimensions of circulation within the block, the extent of outside areas on an estate, or the connections to surrounding spaces, are all now not only issues of design but also of health. The Covid-19 pandemic has taught us that social ritual can overcome confused territoriality and that some un-defensible spaces can be remodelled; Newman's other factors, such as integration into local milieu, have more entrenched impacts. Some design limitations can be overcome (coexisting in segregated spaces is tolerated during Covid), but some (super-density, intensive levels of occupation, height and over-complex circulation patterns) cannot. The six design misperceptions we identified in this chapter show that it is not the principles of defensible space but how they are applied that is a likely predicator of success, and that unsuccessful schemes become inevitably so when the principles are pushed to exaggerated extremes.

At this stage in the global pandemic we can only speculate what a move beyond Newman's defensible space theory might look like, but what is certain is that defensible space will not disappear any time soon as a practical concept, nor as a common sense, middle-range theory.

Notes

1 In Gustafsson's (1983) definition, pseudo-policy lacks the key knowledge needed to implement it. We would argue that knowledge about defensible space is not as much missing, as used in such a fluid way as to undermine its ability to form what Gustafsson would consider real policy rather than symbolic or pseudo-policy.

2 It is worthwhile quoting the GLA's Policy 7.3 'Designing-out-Crime' from the 2016 London Plan in full to show how selected defensible space principles were incorporated, alongside generic recommendations, on permeability into London's planning policy: 'Planning decisions: Developments should reduce the opportunities for criminal behaviour and contribute to a sense of security without being overbearing or intimidating. In particular: a) routes and spaces should be legible and well maintained, providing for convenient movement without compromising security, b) *there should be a clear indication of whether a space is private, semi-public or public, with natural surveillance of publicly accessible spaces* from buildings at their lower floors, c) design should encourage a level of human activity that is appropriate to the location, incorporating a mix of uses where appropriate, to maximise activity throughout the day and night, creating a reduced risk of crime and a sense of safety at all times, d) *places should be well designed to promote an appropriate sense of ownership over communal spaces*, e) *places, buildings and structures should incorporate appropriately designed security features*, f) schemes should be designed to minimise on-going management and future maintenance costs of the particular safety and security measures proposed. The above measures should be incorporated at the design stage to ensure that overall design quality is not compromised' (GLA 2016: 287, emphasis added). The detailed policy refers to the deleted *Safer Places* as well as SBD. However, the latest London Plan 2020 and the associated Estate Regeneration guidance 2018, have deleted all mention of designing-out-crime, focusing instead on optimising density, and working with local communities to make better use of small and existing sites, in an attempt to meet the Mayor's ambitious housing targets.

3 In the Stirling prize-winning social housing scheme Goldsmith Street the 'ginnel' or back alley between the back gardens was 'considered a risk for Norwich Council as unmanaged space' but the generous space, the lower than normal garden fences (and the entry code controlled alley gates at each end) have made them ideal play spaces for children (https://www.ribaj.com/buildings/mikhail-riches-goldsmith-street-social-housing-norwich-passivhaus-stirling-prize).

4 Subsequently re-exporting her version of the concept at many international conferences.

5 The SNU, then the foremost practice-focused crime-reduction research team, were commissioned by the DoE's David Riley, who was simultaneously managing

the DICE evaluation, so it is reasonable to take this quote as reflecting the Department's position.

6   Familiar as Schön's (1983) 'wicked' problems in the swampy lowlands of practical decision-making.

7   Having been neighbourhood manager for Globe Town Neighbourhood during DICE, Steve Stride (mentioned in Chapter 4) became CEO of Poplar HARCA continuing to deliver distinct local neighbourhood focused interventions (https://www.theguardian.com/society-professionals/2016/nov/30/guardian-public-service-awards-2016-transformation-winner-poplar-harca).

8   Unacknowledged ideas pre-existing is similar to Merton's scientific 'pre-discoveries'.

9   And not in the left liberal sense of counter-cultural either.

10 Newman's low opinion of housing professionals' critical facilities might be countered by Marshall's (2012) dream of urban design professionals with the scientific training and skills to interrogate the validity of academic evidence for themselves.

11 For example, the biologist Lamarck who in 1809 developed the concept of evolution as the inheritance of acquired characteristics. At the time this raised controversy and disagreement from clerics and geologists as well as biologists. It was, however, taken seriously and was discussed and refuted within the scientific community. Nonetheless, by 1857 Darwin and Wallace had devised what is accepted as the correct theory of evolution by natural selection.

12 Of course their actions were as much about social engineering as about making good failed housing designs – the 'destruction-proof homes'.

13 *Project news and updates.* https://broadwaterfarmestate.commonplace.is/news.

14 *The Branch Hill Estate, Camden.* https://municipaldreams.wordpress.com/2013/07/02/the-branch-hill-estate-camden-the-most-expensive-council-housing-in-the-world.

15 *A Greater Manchester vision for housing.* https://www.greatermanchester-ca.gov.uk/media/1714/gm_vision_full.pdf.

16 The 2019 draft Greater Manchester Spatial Framework (GMSF) is to be replaced by a new Joint Development Plan Document 'Places for Everyone' after Stockport, one of the 10 Local Authorities covered by the joint DPD, withdrew from the GMSF in 2020 after raising concerns about the potential impacts on the green belt, traffic levels and local services.

17 *International evidence review: High-rise housing as sustainable urban intensification.* https://housingevidence.ac.uk/our-work/research-projects/high-rise-housing-as-sustainable-urban-intensification.

18 *Co-living in London.*https://www.bbc.co.uk/news/uk-england-london-43090849.

19 *Collective living is fine for students but for everybody else it stinks.* https://www.theguardian.com/commentisfree/2016/apr/28/collective-living-students-london-housing-crisis.

20 *Co-living spaces: Modern utopia or over-organised hell?*https://londonist.com/london/housing/co-living-spaces-modern-utopia-or-over-organised-hell.

21 *Too poor to play: children in social housing blocked from communal playground.* https://www.theguardian.com/cities/2019/mar/25/too-poor-to-play-children-in-social-housing-blocked-from-communal-playground; *Segregated playground developer now says all children are welcome.* https://www.theguardian.com/cities/2019/mar/27/segregated-playground-developer-now-says-all-children-are-welcome.

22 *Brokenshire unveils new measures to stamp out poor doors*.https://www.gov.uk/government/news/brokenshire-unveils-new-measures-to-stamp-out-poor-doors.

23 Lees, ethnographic notes from Aylesbury Public Inquiry, 2018.

24 *Inside the tower block refurbished for luxury living.* https://www.ft.com/content/f4e7a2c6-5aa1-11e9-939a-341f5ada9d40; Judith Evans, 'A Brutal Elevation', *Financial Times*, 5 May 2019.

25 *Here's what our cities will look like after the coronavirus pandemic.* https://kinder.rice.edu/urbanedge/2020/03/26/what-our-cities-will-look-after-coronavirus-pandemic.

26 *Cities and public health: Our new challenge in urban planning.* https://www.gensler.com/research-insight/blog/cities-and-public-health-our-new-challenge-in-urban-planning.

27 Covid-19 is reminding us not only of past design solutions or community safety interventions, but the opportunity to revisit a generally more inclusive and equitable cityscape. For example, a 2013 poster created by Justice for Families shows an older woman looking out of her window at a mother and child reading on the front door steps while a young boy cycles past – the strapline 'I don't watch my neighbors, I see them. We make our community safer together' ('A New Way of Understanding "Eyes on the Street"'. https://www.citylab.com/equity/2013/07/new-way-understanding-eyes-street/6276).

28 Photos from Historic England's 'Picturing Lockdown Collection' contain images of chairs pulled out onto the pavement to sunbathe, awnings strung between balconies, children on 'playdates' across garden walls and chalked hopscotch on roads (https://historicengland.org.uk/images-books/archive/collections/photographs/picturing-lockdown).

References

ACPO and Secured by Design. 2004. *Refurbishment of Homes*. Association of Chief Police Officers, London.

ACPO and Secured by Design. 2010. *Secured by Design New Homes 2010*. Association of Chief Police Officers, London. https://www.securedbydesign.com/images/downloads/SBD_New_Homes_2010.pdf (accessed 20 August 2021).

ACPO and Secured by Design. 2014. *Secured by Design New Homes 2014*. Association of Chief Police Officers, London. https://www.securedbydesign.com/images/downloads/New_Homes_2014.pdf (accessed 20 August 2021).

ACPO and Secured by Design. 2016. *Secured by Design Homes 2016: Version 1*. Association of Chief Police Officers, London. https://www.securedbydesign.com/images/downloads/SBD-HOMES-2016.pdf (accessed 20 August 2021).

ACPO and Secured by Design. 2019. *Secured by Design Homes 2019: Version 2*. Association of Chief Police Officers, London. https://www.securedbydesign.com/images/downloads/HOMES_BROCHURE_2019_NEW_version_2.pdf (accessed 20 August 2021).

AEDAS. 2004. *Mozart Handout: Mozart Estate - Building the Vision*. AEDAS Architects, London.

Allen, J. 2004. The whereabouts of power: Politics, government and space. *Geografiska Annaler. Series B, Human Geography* 86, 1, 19–32.

Allen, C., and Imrie, R. (eds.). 2010. *The Knowledge Business*. Ashgate, Aldershot.

Anon. 1985a. Report of speech given by Sir Keith Joseph to the Geographers' Association. *The Geographical Magazine* August. n.p.

Anon. 1985b. Review of 'Utopia on Trial'. *The Alternative Bookshop Broadsheet* July, 1.

Anson, B. 1986. Removing walkways is not nearly enough. *Town and Country Planning* 55, June, 174–175.

Anson, B. 1986. Don't shoot the graffiti man. *Architects' Journal* 184, 27, 16–17.

Defensible Space on the Move: Mobilisation in English Housing Policy and Practice, First Edition. Loretta Lees and Elanor Warwick.
© 2022 Royal Geographical Society (with the Institute of British Geographers). Published 2022 by John Wiley & Sons Ltd.

Anson, B. 1987. Don't shoot the graffiti man. *Architect & Surveyor* 8–9.

Antaki, C., Billig, M., Edwards, D., and Potter, J. 2003. Discourse analysis means doing analysis: A critique of six analytic shortcomings. *Discourse Analysis Online* 1. http://www.shu.ac.uk/daol/articles/v1/n1/a1/antaki2002002-t.html (accessed 20 August 2021).

Architects' Journal. 1985. Coleman's utopia goes on trial. 182, 51/52, 8.

Architects' Journal. 1986. Coleman rules ok? 183, 4, 25.

Architects' Journal. 1990. Colemanisation goes on trial. 192, 4, 9.

Ardill, J. 1986. Ministers bow to vision of utopia for cities. *The Guardian* 17 March, n.p.

Armitage, R. 2000. *An Evaluation of Secured by Design Housing within West Yorkshire – Briefing Note 7/00*. Home Office, London.

Armitage, R. 2004. Secured by design - An investigation of its history, development and future role in crime reduction. PhD thesis, University of Huddersfield.

Armitage, R. 2013. *Crime Prevention through Housing Design: Policy and Practice*. Palgrave Macmillan, Basingstoke.

Armitage, R., Colquhoun, I., Ekblom, P., Monchuk, L., Pease, K., and Rogerson, M. 2010. *Residential Crime and Design: Final Report for CABE and HO*. Applied Criminology Centre University of Huddersfield, Huddersfield.

Armitage, R., and Ekblom, P. (eds.). 2019. *Rebuilding Crime Prevention through Environmental Design: Strengthening the Links with Crime Science*. Routledge, London.

Armitage, R., and Monchuk, L. 2011. Sustaining the crime reduction impact of designing out crime: Re-evaluating the Secured by Design scheme 10 years on. *Security Journal* 24, 320–343.

Armstrong, J. 1985. Doctrinaire gimmickry denied peoples' dreams. *Town and Country Planning* 54, 7, July/August, 242.

Ash, J. 1985. Polemic with statistics. *Building Design* 765, 18 October, 38.

Ash, M., Burbridge, M., Hillman, J., Hollamby, E., Law, S., Lyons, E., and Newman, O. 1975. *Architecture, Planning and Urban Crime*. National Association for the Care and Resettlement of Offenders (NACRO), London.

Atlas, R. 1991. The other side of CPTED. *Architecture and Security* March, 63–66.

Audit Commission. 1986. *Managing the Crisis in Council Housing*. Audit Commission for Local Authorities in England and Wales, London. http://webarchive.nationalarchives. gov.uk/+/http:/www.hm-treasury.gov.uk/consultations_and_legislation/barker/consult_barker_index.cfm (accessed 20 August 2021).

Bailey, N., and Manzi, T. 2008. *Developing and Sustaining Mixed Tenure Housing Developments*. JRF, York.

Baillieu, A. 1991a. Planning system calls in police. *Building Design* 1052, 11 October, 1.

Baillieu, A. 1991b. Unlucky throw for DICE projects. *Building Design* 1052, 11 October, 1.

Baker, T., and McCann, E. 2018. Beyond failure: The generative effects of unsuccessful proposals for Supervised Drug Consumption Sites (SCS) in Melbourne, Australia. *Urban Geography* 41, 9, 1179–1197.

Baker, T., and McGuirk, P. 2017. Assemblage thinking as methodology: Commitments and practices for critical policy research. *Territory, Politics, Governance* 5, 425–442.

Baker, T., and Walker, C. (eds.). 2019. *Public Policy Circulation: Arenas, Agents and Actions*. Edward Elgar, London.

Baldassare, M. 1975. Review: 'Human identity in the urban environment, defensible space: Crime prevention through urban design'. *Contemporary Sociology* 4, 4, 435–436.

Banham, R. 1973. Review 'Defensible space'. *New Society* 66, 18 October, 154–156.

Bar-Hillel, M. 1986a. Coleman's progress. *Building Design* 813, 21 November, 14.

Bar-Hillel, M. 1986b. Boost for scourge of the ghettos. *Chartered Surveyor's Weekly* 27 April, n.p.

Bar-Hillel, M. 1986c. Coleman slams DoE crime package. *Chartered Surveyor's Weekly* 11 December, n.p

Bar-Hillel, M. 1997. Cut back on parking spaces to squeeze in more housing. *London Evening Standard* 14 November, 15.

Bar-Hillel, M. 1998. Letter: Road rage. *Building Design* 1423, 20 November, 10.

Bar-Hillel, M. 2014. Unsafe as houses. Review of *All that is Solid: The great housing disaster* by Danny Dorling. *Literary Review* 420, May, 46.

Barker, K. 2004. *Delivering Stability: Securing Our Future Housing Needs*. HM Treasury, London. https://www.thenbs.com/PublicationIndex/documents/details?Pub=HMT&DocID=265761 (accessed 20 August 2021).

Barnes, T. 2004a. Placing ideas: Genius loci, heterotopia and geography's quantitative revolution. *Progress in Human Geography* 28, 5, 565–595.

Barnes, T. 2004b. A paper related to everything, but more related to local things. *Annals of the Association of American Geographers* 94, 278–283.

Barnes, T., and Sheppard, E. (eds.). 2019. *Spatial Histories of Radical Geography: North America and Beyond*. Wiley, Chichester.

Baxter, R., and Lees, L. 2009. The rebirth of high-rise living in London: Towards a sustainable and liveable urban form, in Imrie, R., Lees, L., and Raco, M. (eds.), *Regenerating London: Governance, Sustainability and Community in a Global City*. Routledge, London, pp. 151–172.

Beharrell, A. 2017. Altered estates. *Urban Design Group Journal* 143, 14–18. https://www.udg.org.uk/sites/default/files/publications/UD143%20Urban%20Design%20-%20Estate%20Regeneration.pdf (accessed 20 August 2021) (accessed 20 August 2021).

Bellush, J., and Netzer, D. 1990. *Urban Politics: New York Style*. M. E. Sharpe, New York.

Benson, D. 2009. Review article: Constraints on policy transfer. CSERGE Working Paper EDM, No. 09-13. The Centre for Social and Economic Research on the Global Environment (CSERGE), University of East Anglia.

Berg, B. 2007. *New York City Politics: Governing Gotham*. Rutgers University Press, New Brunswisk, NJ.

Birkbeck, D., and Kruczkowski, S. 2020. *Building for Healthy Life* https://www.designforhomes.org/project/building-for-life (accessed 20 August 2021).

Birtchnell, J., Masters, N., and Deahl, M. 1988. Depression and the physical environment: A study of young women on a London housing estate. *British Journal of Psychiatry* 153, 56–64.

Blanc, F., Scanlon, K., and White, T. 2020. *Living in a denser London: How residents see their homes*, LSE, London. https://www.lse.ac.uk/geography-and-environment/research/lse-london/documents/Reports/2020-LSE-Density-Report-digital.pdf (accessed 20 August 2021).

Blandy, S. 2007. Gated communities in England as a response to crime and disorder: Context, effectiveness and implications. *People, Place & Policy* 1/2, 47–54.

Blomley, N. 2004. Un-Real estate: Propriety space and public gardening. *Antipode* 36, 4, 614–639.

Bok, R. 2014. Airports on the move? The policy mobilities of Singapore Changi Airport at home and abroad. *Urban Studies* 52, 14, 2724–2740.

Booker, C. 1985. The chilling legacy of Utopia. *The Daily Mail* n.p.

Bottoms, A.E. 1974. Review of 'Defensible space'. *The British Journal of Criminology* 14, 2, 203–206.

Boughton, J. 2018. *Municipal Dreams: The Rise and Fall of Council Housing*. Verso London.

Boys Smith, N., and Morton, A. 2013. *Create Streets: Not just Multi-storey Estates*, Policy Exchange https://policyexchange.org.uk/wp-content/uploads/2016/09/create-streets.pdf (accessed 20 August 2021).

Boys Smith, N., and Wildblood, J. 2014. *Alice in Wonderland* https://www.createstreets.com/wp-content/uploads/2017/12/Interview-with-Alice-Coleman.pdf (accessed 20 August 2021).

Brecher, C., Horton, R., Cropf, R., and Mead, D.M. 1993. *Power Failure: New York City Politics and Policy Since 1960*. Oxford University Press, New York.

Brenner, N., and Theodore, N. 2002. Cities and the geographies of 'actually existing neoliberalism'. *Antipode* 34, 3, 349–379.

Brindley, T., Rydin, Y., and Stoker, G. 1996. *Remaking Planning: The Politics of Urban Change*. Routledge, London.

Broady, M. 1966. Social theory in architectural design. *Arena: The Architectural Association Journal* 81, 149–154.

Brown, J. 1974. High rise flats are nurseries of crime. *Sheffield Morning Telegraph* 4 May, n.p.

Brown, J. 1999. An evaluation of the Secured by Design initiative in Gwent, South Wales. MSc dissertation. University of Leicester, Leicester.

Brown, K. 2004. Genderism and the bathroom problem: (Re)materialising sexed sites, (re)creating sexed bodies. *Gender, Place and Culture* 11, 331–346.

BSI. 1986. *BS 8220-1:1986 Guide for Security of Buildings against Crime. Part 1: Dwellings*. British Standards Institution (BSI), Milton Keynes.

BSI. 2000. *BS 8220-1:2000 Guide for Security of Buildings against Crime. Part 1: Dwellings*, 2nd edn. British Standards Institution (BSI), Milton Keynes.

Building Design. 1973. Review of 'Defensible space'. (August 8) 114, 3.

Building Design. 1985a. Preview of World in Action television programme. (November 15) 765, n.p.

Building Design. 1985b. [Editorial comments]. (August 2) 740, 2.

Building Design. 1985c. Review of 'When we build again: Let's have housing that works' by Colin Ward. (September 6) 755, 17.

Building Design. 1986. Patten hard on crime. (March 4) 778, 14.

Building Design. 1988. Coleman's ideas win backing. (January 1) 870, 1.

Building Design. 1997. Is this the future? (November 14) 1370, 1.

Building Design. 1998. Coleman can't cut the mustard. (November 7) 1422, 12.

Bulos, M. 1987. Housing management: New ideas, new approaches. *High Rise Quarterly* 2, 1, 11–12.

Bulos, M., and Walker, S. 1987. British high rise housing in context, in Bulos, M. and Walker, S. (eds.), *The Legacy and Opportunity for High Rise Housing in Europe: The Management of Innovation.* Housing Studies Group, London, pp. 5–22.

Burawoy, M. 2004. Public sociologies: Contradictions, dilemmas and possibilities. *Social Forces* 82, 1–16.

Burawoy, M. 2005. The critical turn to public sociology. *Critical Sociology* 31, 313–326.

Burbidge, M. 1973. *Vandalism: A Constructive Approach.* Department of the Environment, London.

Burbidge, M., Kirby, K., and Wilson, S. 1980. *An Investigation of Difficult to Let Housing.* Department of the Environment, London.

CABE. 2008. Minutes of meeting between CABE/ACPO 20 September [unpublished] Commission for Architecture and the Built Environment (CABE).

CABE. 2009. *This Way to Better Residential Streets.* Commission for Architecture and the Built Environment (CABE), London.

CABE. 2010. *Mapping Space Standards for the Home.* Commission for Architecture and the Built Environment (CABE), London. https://webarchive.nationalarchives. gov.uk/ukgwa/20110118111539mp_/http://www.cabe.org.uk/files/mapping-existing-housing-standards.pdf (accessed 20 August 2021).

CABE. 2014. *Creating Safe Places to Live through Design.* Design Council CABE, London. http://www.designcouncil.org.uk/knowledge-resources/report/creating-safe-places-live-through-design (accessed 20 August 2021).

Cairney, P. 2011. The new British policy style: From a British to a Scottish political tradition? *Political Studies Review* 9, 2, 208–220.

Cameron, D. 2016. Cameron: I will bulldoze sink estates. *The Sunday Times*, 10 January. https://www.thetimes.co.uk/article/cameron-i-will-bulldoze-sink-estates-2qzmgkgwlxn (accessed 20 August 2021).

Cameron, S. 1987. Inner-city in fighting. *The Times* 20 July.

Campkin, B. 2013. *Remaking London: Decline and Regeneration in Urban Culture.* I.B. Tauris, London/New York.

Carmona, M. 2012. The London way: The politics of London's strategic design. *Architectural Design* 82, 36–43.

Carmona, M. 2013. By design – Bye bye. *Design Matters* 37. https://matthew-carmona. com/2013/01/14/by-design-bye-bye/ (accessed 20 August 2021).

Carmona, M., De Magalhaes, C., and Natarajan, L. 2017. Design governance the CABE way, its effectiveness and legitimacy. *Journal of Urbanism: International Research on Placemaking and Urban Sustainability* 11, 1, 1–23.

Castree, N. 2006. Geography's new public intellectuals. *Antipode* 38, 396–412.

Centre for Cities. 2019. *Cities Outlook 2019: A Decade of Austerity*. Centre for Cities, London. https://www.centreforcities.org/reader/cities-outlook-2019/a-decade-of-austerity (accessed 20 August 2021).

Centre for London. 2016. *Another Storey: The Real Potential for Estate Densification*. Centre for London, London. https://www.centreforlondon.org/publication/estate-densification (accessed 20 August 2021).

Chakrabortty, A., and Robinson-Tillett, S. 2014. The truth about gentrification: Regeneration or con trick? *The Guardian* 18 May 2014.

Chamberlain, S., and Leydesdorff, S. 2004. Transnational families: Memories and narratives. *Global Networks* 4, 3, 227–241.

Chartered Institute of Housing. 1989. *Housing*. CIH, London.

Chou, W., and Dancygier, R. 2021. Why parties displace their voters: Gentrification, coalitional change, and the demise of public housing. *American Political Science Review* 115, 2, 429–449.

Cisneros, H.G. 1996. Defensible space: Deterring crime and building community. *Cityscape: A Journal of Policy Development and Research* 49, December. http://www.huduser.org/Periodicals/CITYSCPE/SPISSUE/ch2.pdf (accessed 20 August 2021).

City of Towers. 1977. [TV programme]. BBC, BBC 2. Directed by Booker, C. Available at: http://www.youtube.com/watch?v=EY_YuXf8ZU8 (accessed 4 September 2014).

City of Westminster Council. 1993. *Director of Housing's Report to Housing Committee: Mozart Estate – Estates Action Bid*. City of Westminster Council, London.

CityWest Homes. 2012. *Mozart Estate Newsletters*. Spring 2011/Winter 2012, CityWest Homes, London. http://www.cwh.org.uk/locations/mozart/local-newsletters (accessed 4 September 2014).

Clarke, J., Bainton, D., Lendvai, N., and Stubbs, P. 2015. *Making Policy Move: Towards a Politics of Translation and Assemblage*. Policy Press, Bristol.

Clarke, A., and Dawson, R. 1999. *Evaluation Research: An Introduction to Principles, Methods and Practice*. Sage, London.

Clarke, N. 2012. Urban policy mobility, anti-politics, and histories of the transnational municipal movement. *Progress in Human Geography* 36, 25–43.

Cohen, L., and Felson, M. 1979. Social change and crime rate trends: A routine activity approach. *American Sociological Review* 44, 4, 588–608.

Cohen, L., Manion, M., and Morrison, M. 2018. *Research Methods in Education*. Routledge, London.

Cole, I., and Furbey, R. 1994. *The Eclipse of Council Housing*. Routledge, London.

Coleman, A. 1980. The death of the inner city: Cause and cure. *London Journal* 6, 1, 3–22.

Coleman, A. 1984a. Trouble in Utopia: Design influences in blocks of flats. *The Geographical Journal* 150, 3, 351–358.

Coleman, A. 1984b. Planner's crimes: Dreams that went sour. *The Geographical Magazine* 560–565.

Coleman, A. 1985a. *Utopia on Trial: Vision and Reality in Planned Housing*. Hilary Shipman, London.

Coleman, A. 1985b. Indefensible space. *Building Design* 738, 5 October, 14.

Coleman, A. 1985c. Refurbishing the housing stock: Mozart estate. *RIBA Journal* 92, June, 50–52.

Coleman, A. 1986a. Design improvement: Utopia goes on trial. *Town and Country Planning* 55, 5, 138–140.

Coleman, A. 1986b. Utopia debate. *Architects' Journal* 184, 32, 16–17.

Coleman, A. 1988. Home is home. *Architects' Journal* 187, 43, 31.

Coleman, A. 1989. Design improved control experiment. *High Rise Quarterly* 3, 3, 4–5.

Coleman, A. 1990. *Utopia on Trial: Vision and Reality in Planned Housing*, 2nd edn. Hilary Shipman, London.

Coleman, A. 1996. Letter to the Editor: Three-strikes policy. *The Times*, 15 October, n.p.

Coleman, A. 2009. The psychology of housing. *The Salisbury Review: The Quarterly Magazine of Conservative Thought* 28, 2, 10–12.

Coleman, A. 2013a. The curse of modernism. [unpublished].

Coleman, A. 2013b. The planning travesty: Part 1. [unpublished].

Coleman, A., and Brown, S. 1985. *Report to A2(3) Branch Metropolitan Police. London.* DICE Design Research Unit, King's College London, London.

Coleman, A., and Cross, D. 1994. *Rogers Estate: Design Improvement Report*. DICE Design Research Unit, King's College London, London.

Coleman, A., and Cross, D. 1995. A cross-national comparison of harmful housing design: DICE design potential in Canada. *Land Use Policy* 12, 2, 145–164.

Coleman, A., Cross, D., and Silver, P. 1994. *Ranwell East Estate: Design Improvement Report*. DICE Design Research Unit, King's College London, London.

Coleman, A., Cross, D., and Silver, P. 1995. DICE design guide for new-build houses Mozart estate. DICE Design Research Unit, King's College London, London.

Coleman, A., Faith, P., Cross, D., Silver, P., Harvey, A., and Furtado, R. 1992. *The Mozart Estate, Design Improvement Proposals Stage III*. DICE Consultancy, Kings College London, London.

Coleman, A., and Naylor, P. 1993. *Ranwell East Estate: Design Improvement Report*. DICE Design Research Unit, King's College London, London.

Coleman, A., and Silver, P. 1995a. *Methods of Estate Improvement: DICE Consultancy Leaflet No 4*. DICE Design Research Unit, King's College London, London.

Coleman, A., and Silver, P. 1995b. *The DICE Project: DICE Leaflet No 5. London.* DICE Design Research Unit, King's College London, London.

Cook, B. 1977. *A Look at Vandalism in the U.S.A.* Central Office of Information. HMSO, London.

Cook, I. 2008. Mobilising urban policies: The policy transfer of US business improvement districts to England and Wales. *Urban Studies* 45, 773–795.

Cook, I. 2015. Policy mobilities and interdisciplinary engagement. *International Journal of Urban and Regional Research* 39, 835–837.

Cooper, I. 2006. *Sustainable Construction and Planning: The Policy Agenda. LSE SusCon Project.* LSE Centre for Environmental Policy and Governance, London.

Cowan, R. 2005. *The Dictionary of Urbanism.* Streetwise Press, Tisbury, UK.

Cozens, P. 2005. *Designing Out Crime. From Evidence to Action. Delivering Crime Prevention.* Australian Institute of Criminology, Sydney.

Cozens, P. 2015. Crime and community safety: Challenging the design consensus, in Barton, H., Thompson, S., Grant, M., and Burgess, S. (eds.), *Routledge Handbook of Planning for Health and Well-being.* Routledge, London, pp. 162–177.

Cozens, P. 2016. *Think Crime! Using Evidence, Theory and Crime Prevention through Environmental Design (CPTED) for Planning Safer Cities,* 2nd edn. Praxis Education, Quinn's Rock Perth, WA.

Cozens, P., Hillier, D., and Prescott, G. 2001a. Crime and the design of residential property – exploring the perceptions of planning professionals, burglars and other users: Part 2. *Property Management* 19, 4, 222–248.

Cozens, P., Hillier, D., and Prescott, G. 2001b. Crime and the design of residential property – exploring the theoretical background: Part 1. *Property Management* 19, 2, 136–164.

Cozens, P., and Love, T. 2017. The dark side of crime prevention through environmental design (CPTED). *Criminology and Criminal Justice,* https://oxfordre.com/criminology/view/10.1093/acrefore/9780190264079.001.0001/acrefore-9780190264079-e-2 (accessed 20 August 2021).

Cozens, P., Saville, G., and Hillier, D. 2005. Crime prevention through environmental design (CPTED): A review and modern bibliography. *Property Management* 23, 5, 328–356.

Craggs, R., and Neate, H. 2017. Post-colonial careering and urban policy mobility: Between Britain and Nigeria, 1945–1990. *Transactions of the Institute of British Geographers* 42, 44–57.

Crawford, A., and Jones, M. 1996. Kirkholt revisited: Some reflections on the transferability of crime preventions. *The Howard Journal of Crime and Justice* 35, 1, 21–39.

Cross, M., and Keith, M. (eds.). 1993. *Racism, the City and the State.* Routledge, London.

Crossman, R. 1975. *Diaries of a Cabinet Minister: Volume 1 Minister of Housing 1964–1966.* Hamish Hamilton/Jonathan Cape, London.

Cupers, K. 2016. Human territoriality and the downfall of public housing. *Public Culture* 29, 1, 165–190.

Davidson, M., and Lees, L. 2010. New-build gentrification: Its histories, trajectories, and critical geographies. *Population, Space and Place* 16, 395–411.

Davidson, R. 1981. *Crime and Environment.* Croom Helm, London.

Davie, M. 1984. We were all wets when we were building Utopia: Interview with Sir Hugh Casson. *The Observer* 2 December, n.p.

Davie, M. 1986. It's hell for tenants in the bureaucrats' Utopia. *The Observer,* n.p.

Davies, G. 2017. The Grenfell disaster shows that architecture is always political. https://failedarchitecture.com/the-grenfell-disaster-shows-that-architecture-is-always-political-political-with-a-capital-p (accessed 20 August 2021).

DCLG. 2012. *National Planning Policy Framework.* Department for Communities and Local Government, London. https://www.gov.uk/government/uploads/system/uploads/attachment_data/file/6077/2116950.pdf (accessed 20 August 2021).

DCLG. 2014a. *Housing Standards Review: Summary of Responses.* Department for Communities and Local Government, London. https://www.gov.uk/government/uploads/system/uploads/attachment_data/file/289144/140225_final_hsr_summary_of_responses.pdf (accessed 20 August 2021).

DCLG. 2014b. *Live Tables. Table 254 Housebuilding: Permanent Dwellings Completed, by House and Flat, Number of Bedroom and Tenure, England.* Department of Communities and Local Government, London. https://www.gov.uk/government/statistical-data-sets/live-tables-on-house-building (accessed 20 August 2021).

DCLG. 2014c. *Estate Regeneration Programme: Prospectus for 2015/16.* https://assets.publishing.service.gov.uk/government/uploads/system/uploads/attachment_data/file/320196/140613_Estate_Regeneration_Prospectus.pdf (accessed 20 August 2021).

DCLG. 2016. *Estate Regeneration National Strategy,* https://www.gov.uk/guidance/estate-regeneration-national-strategy (accessed 20 August 2021).

DeLanda, M. 2016. *Assemblage Theory.* Edinburgh University Press, Edinburgh.

DETR and CABE. 2000. *By Design: Urban Design in the Planning System: Towards Better Practice.* Thomas Telford, Tonbridge.

DETR and CABE. 2001. *By Design: Better Places to Live: A Comparison Guide to PPG3.* Thomas Telford, Tonbridge.

DfT and DCLG. 2007. *Manual for Streets.* Department for Transport and Department for Communities and Local Government, London.

Dodd, N. 1994. *The Sociology of Money.* Polity Press, Cambridge.

DoE. 1972. *Design Bulletin 25: The Estate Outside the Dwelling.* Department of the Environment, London.

DoE. 1976. *Research Report 6: The Value of Standards for the External Residential Environment.* Department of the Environment, London.

DoE. 1981. *Security on Council Estates.* Department of the Environment, London.

DoE. 1984. *More Than Just a Road.* Department of the Environment, London.

DoE. 1986. *Urban Housing Renewal Unit Annual Report 1985–86.* HMSO, London.

DoE. 1989. *Handbook of Estates Improvement.* Department of the Environment, London.

DoE. 1991. *The Estates Action Handbook.* Department of the Environment, London.

DoE. 1994. *Planning Out Crime: Circular 5/94.* Department of the Environment, London.

DoE. 1996. *An Evaluation of Six Early Estate Action Schemes.* HMSO, London.

Doig, B., Littlewood, J., and Mason, S. 1992. *Policy Evaluation: The Role of Social Research.* Department of the Environment, London.

Dolowitz, D. 1998. *Learning from America: Policy Transfer and the Development of the British Workfare State.* Sussex University Press, Brighton.

Dolowitz, D., and Marsh, D. 1996. Who learns what from whom: A review of the policy transfer literature. *Political Studies* 44, 343–357.

Dolowitz, D., and Marsh, D. 2000. Learning from abroad: The role of policy transfer in contemporary policy-making. *Governance* 13, 1, 5–23.

Dolowitz, D., and Marsh, D. 2012. The future of policy transfer research. *Political Studies Review* 10, 339–345.

Domosh, M. 1991. Toward a feminist historiography of geography. *Transactions of the Institute of British Geographers* 16, 1, 95–104.

Dorling, D. 2014. *All that Is Solid: How the Great Housing Disaster Defines Our Times and What We Can Do about It*. Allen Lane, London.

Downes, D. 1974. Review: *Vandalism* edited by Colin Ward. *The British Journal of Criminology* n.p.

Downes, D., and Rock, P. 2007. *Understanding Deviance: A Guide to the Sociology of Crime and Rule Breaking*. Oxford University Press, Oxford.

Dunleavy, P. 1981. *The Politics of Mass Housing in Britain, 1945–1975*. Clarendon Press, Oxford.

Dunne, J., and Davenport, J. 2012. Police seize 'drug gang leaders' in warning to stay away from Games. *Evening Standard* 25 July, n.p.

Dussauge-Laguna, M. 2012. On the past and future of policy transfer research: Benson and Jordan revisited. *Political Studies Review* 10, 313–324.

Economist. 2006. Field of dreams. *The Economist*. 27 July http://www.economist.com/node/7226156 (accessed 20 August 2021).

Edmunds, L. 1984. Doing the ground works to improve inner-city living. *The Daily Telegraph* 26 April, 16–17.

Edwards, M., and Evans, M. 2011. *Getting Evidence into Policy-making*. ANSOG Institute for Governance, University of Canberra, Canberra.

Ekblom, P. 2011. Deconstructing CPTED … and reconstructing it for practice, knowledge management and research. *European Journal on Criminal Policy and Research* 1, 17, 7–28.

Essex Design Guide. 2018. Private space. https://www.essexdesignguide.co.uk/design-details/layout-details/private-space (accessed 20 August 2021).

Evans, J. 2016. Trials and tribulations: Problematizing the city through/as urban experimentation. *Geography Compass* 10, 429–443.

Evening Standard. 2012. Eight arrested in west London Mozart Estate dawn swoop to smash gang's violent drug war. *London Evening Standard* 2 May.

Farr, J., and Osborn, S. 1997. *High Hopes: Concierge, Controlled Entry and Similar Schemes for High Rise Blocks*. DoE, London.

Fenton, A. 2012. Reduced statistics: Housing and communities in England, in Cohen, R., Greig, A., and Brownstein, L. (eds.), *Mis-measurement of Health and Wealth*. Radical Statistics Group, London. https://www.radstats.org.uk/no107/Fenton107.pdf (accessed 20 August 2021).

Ferreri, M. 2020. Painted bullet holes and broken promises: Understanding and challenging municipal dispossession in London's public housing 'decanting'. *International Journal of Urban and Regional Research* 44, 6, 1007–1022.

Fischer, F. 1995. Citizens and experts in the public sphere: From technocratic politics to participatory deliberation. *Indian Journal of Public Administration* 41, 3, 525–533.

Floyd Slaski Partnership. 1993. *Mozart Estate: Form B Programme for Regeneration, Estate Action Submission for Westminster City Council.* [Form B proforma and report]. Floyd Slaski Partnership, London.

Flyvbjerg, B. 1998. *Rationality and Power: Democracy in Practice.* The University of Chicago Press, Chicago and London.

Flyvbjerg, B. 2001. *Making Social Science Matter: Why Social Enquiry Fails and How It Can Succeed Again.* Cambridge University Press, Cambridge.

Forrest, R., and Murie, A. 1988. *Selling the Welfare State: The Privatization of Public Housing.* Routledge, London.

Forscher, B. 1963. Chaos in the brickyard. *Science* 142, 18 October, 339–341.

Foster, J., and Hope, T. 1993. *Housing, Community and Crime: The Impact of the Priority Estates Project: Home Office Research Study 131.* Home Office, London.

Franks, A. 1990. Creating close harmony down on the Mozart Estate. *The Times* 21 August, n.p.

Freedman, A. 1987. Public housing: Among experts, modern is a dirty word. *The Globe and Mail (Canada) October,* 3.

Freeman, R. 2012. Reverb: Policy making in wave form. *Environment and Planning A* 44, 1, 13–20.

Friedmann, J. 1973. Review: The human dimension of housing: 'defensible space: crime prevention through urban design' by Oscar Newman. *Growth and Change* 4, 4, 49.

Fuller, D. 2008. Public geographies: Taking stock. *Progress in Human Geography* 32, 834–844.

Fyfe, N., Bannister, J., and Kearns, A. 2006. (In)civility and the city. *Urban Studies* 43, 5/6, 853–861.

Gamman, L., and Thorpe, A. 2014. Design for democratic crime prevention, in Campkin, B. and Ross, R. (eds.), *Design and Trust,* Urban Pamphleteer No. 3. University College London London.

Gilroy, P. 1987. *There Ain't No Black in the Union Jack: The Cultural Politics of Race and Nation.* Unwin Hyman, Chicago.

GLA. 2004. *London Plan.* https://www.london.gov.uk/sites/default/files/the_london_plan_2004.pdf (accessed 20 August 2021).

GLA. 2008. *Supplementary Planning Guidance Providing for Children and Young People's Play and Informal Recreation.* https://www.london.gov.uk/sites/default/files/gla_migrate_files_destination/spg-2008-children-recreation.pdf (accessed 21 August 2021).

GLA. 2010. *London Housing Design Guide.* https://www.london.gov.uk/sites/default/files/interim_london_housing_design_guide.pdf (accessed 21 August 2021).

GLA. 2014a. *Draft Further Alterations to the London Plan: The Spatial Development Strategy For Greater London.* Greater London Authority, London. https://www.london.gov.uk/priorities/planning/london-plan/draft-further-alterations-to-the-london-plan (accessed 20 August 2021).

GLA. 2014b. *Homes for London.* https://www.london.gov.uk/sites/default/files/draft_london_housing_strategy_april_2014_0.pdf (accessed 20 August 2021).

GLA. 2016. *Homes for Londoners: Draft Good Practice Guide to Estate Regeneration.* https://www.london.gov.uk/sites/default/files/draftgoodpracticeestateregenerationguidedec16v2.pdf (accessed 20 August 2021).

GLA. 2018. *Better homes for local people.* https://www.london.gov.uk/sites/default/files/better-homes-for-local-people-the-mayors-good-practice-guide-to-estate-regeneration.pdf (accessed 20 August 2021).

GLA. 2021a. *Good Quality Homes for All Londoners Guidance (consultation draft).* https://consult.london.gov.uk/good-quality-homes-for-all-londoners (accessed 20 August 2021).

GLA. 2021b. *The London Plan.* https://www.london.gov.uk/what-we-do/planning/london-plan/new-london-plan/london-plan-2021 (accessed 20 August 2021).

Glancy, J. 1998. End of the road for Cul-de-sacs. *The Guardian* 27 November, 16.

Glennerston, H., and Turner, T. 1993. *Estate Based Management: An Evaluation.* DoE, London.

Gonzalez, S. 2011. Bilbao and Barcelona 'in motion'. How urban regeneration 'models' travel and mutate in the global flows of policy tourism. *Urban Studies* 48, 7, 1397–1418.

Gorst, T. 1986. Alice in Blunderland. *Building Design* 770, 17 January, 2.

Gosselin, C. 2016. Trans. Waine, O. Urban renewal and the 'Defensible Space' Model: The growing impact of security issues on the way our cities develop. https://www.metropolitiques.eu/Urban-renewal-and-the-defensible.html (accessed 20 August 2021).

Government Social Research Unit. 2003. *The Magenta Book: Guidance for Evaluation,* 1st edn. HM Treasury, London.

Green, S., and Schweber, L. 2008. Theorizing in the context of professional practice: The case for middle-range theories. *Building Research and Information* 36, 6, 649–654.

Guest, P. 1990. Utopia goes on trial. *Building* 255, 20–22.

Guest, P. 1993. Alice proves wonders of looking through glass. *Building* 258, 18.

Gulson, K., Lewis, S., Lingard, B., Lubienski, C., Takayama, K., and Taylor Webb, P. 2017. Policy mobilities and methodology: A proposition for inventive methods in education policy studies. *Critical Studies in Education* 58, 2, 224–241.

Gustafsson, G. 1983. Symbolic and pseudo policies as responses to diffusion of power. *Policy Sciences* 15, 3, 269–287.

Habraken, N. 1998. *Structure of the Ordinary.* MIT Press, Cambridge and London.

Hackney, R. 1987. Building communities, in Sneddon, J. and Theobald, C. (eds.), *Building Communities.The First International Conference of Community Architecture, Planning and Design,* 26–29 November. Community Architecture Information Services (CAIS), Astoria Theatre, London, p. vii.

Hall, S., Critcher, C., Jefferson, T., Clarke, J., and Roberts, B. 1978. *Policing the Crisis: Mugging, the State, and Law and Order.* Macmillan, London.

Hall, S., Murie, A., Rowlands, R., and Sankey, S.. 2004. *Restate: Restructuring Large Housing Estates in European Cities: Good Practices and New Visions for Sustainable Neighbourhoods and Cities.* Report 3j: Large Housing Estates in United Kingdom, Policies and practices. Urban and Regional Research Centre, University of Utrecht, Utrecht.

Halpern, D. 1995. *Mental Health and the Built Environment: More than Bricks and Mortar?* Taylor and Francis, London.

Hanley, L. 2007. *Estates: An Intimate History.* Granta Books, London.

Harman, J. 2012. *Viability Testing Local Plans: Advice for Planning Practitioners.* HBF/ NHBC, London. https://www.local.gov.uk/sites/default/files/documents/viability-testing-local-p-42b.pdf (accessed 20 August 2021).

Harrison, T. 2000. Urban policy: Addressing wicked problems, in Davies, H., Nutley, S., and Smith, P. (eds.), *What Works? Evidence Based Policy and Practice in Public Services.* The Policy Press, Bristol, pp. 207–227.

Harvey, D. 1989. From managerialism to entrepreneurialism: The transformation of urban governance in late capitalism. *Geografiska Annaler. Series B, Human Geography* 71, 1, 3–17.

Harvey, D. 1997. *Justice, Nature and the Geography of Difference.* Wiley-Blackwell, Chichester.

Harvey, D. 2001. The condition of postmodernity, in Seidman, S. and Alexander, J. (eds.), *The New Social Theory Reader: Contemporary Debates.* Routledge, pp. 176–184.

Harvey, D. 2007. *A Brief History of Neoliberalism.* Oxford University Press, Oxford.

Hatherley, O. 2015. Counting broken windows. *The White Review,* November. http://www.thewhitereview.org/white_screen/counting-broken-windows (accessed 20 August 2021).

Hawkes, N. 1991. Architects try to design crime out of the social-breakdown estate. *The Times* 29 August, n.p.

HCA. 2007. *Design and Quality Standards.* Homes and Communities Agency, London. http://www.homesandcommunities.co.uk/sites/default/files/our-work/design_quality_standards.pdf (accessed 20 August 2021).

HC Deb, 17 October 2002, col.943W http://www.publications.parliament.uk/pa/cm200102/cmhansrd/vo021017/text/21017w10.htm#21017w10.html_wqn13 (accessed 20 August 2021).

HC Deb, 21 January 2013, col.76W http://www.publications.parliament.uk/pa/cm201213/cmhansrd/cm130121/text/130121w0004.htm (accessed 20 August 2021).

Healey, P. 2010. Transnational flow of planning ideas and practices. *Royal Geographical Society – Institute of British Geographers Conference 2010.* London, 1–3 September.

Heaven, B. 1986. Comeback on Coleman. *Architects' Journal* 164, 36, 32–33.

Heck, S. 1987. Oscar Newman Revisited. *Architects' Journal* 185, 14, 8 April, 30.

Herbert, D. 1982. *The Geography of Urban Crime.* Longman Inc, New York.

Herbert, D. 1986. First steps to a new science. *Town and Country Planning* 55, 5, 144–145.

Hill, M. 1997. Implementation theory: Yesterdays issue? *Policy and Politics* 25, 375–385.

Hillier, B. 1973. In defense of space. *RIBA Journal* 80, 11, 539–544.

Hillier, B. 1986a. City of Alice's dreams. *Architects' Journal* 184, 28, 39–41.

Hillier, B. 1986b. Letter: Hard science or hard sell? *Architects' Journal* 184, 33, 14.

Hillier, B. 1988. Against enclosure, in Teymur, N., Markus, T.A., and Woolley, T. (eds.), *Rehumanizing Housing.* Butterworths, Oxford, pp. 63–85.

Hillier, B. 1998. The word on the street. *Building Design* 1424, 27 November, 6.

Hillier, B., and Penn, A. 1986a. *The Mozart Estate Redesign Proposals*. Unit for Architectural Studies, University College London, London.

Hillier, B., and Penn, A. 1986b. Mozart plan out of tune says report. *Architects' Journal* 18.

Hillier, B., and Sahbaz, O. 2007. Beyond hot spots: Using space syntax to understand dispersed patterns of crime risk in the built environment. *Crime Hotspots: Behavioural Computation and Mathematical Models Conference*. Institute for Pure and Applied Mathematics, University of California, Los Angeles. 29 January–2 February.

Hillier, B., and Sahbaz, O. 2008. *An Evidence-based Approach to Crime and Urban Design. Or, Can We Have Vitality, Sustainability and Security all at Once?* http://spacesyntax.com/wp-content/uploads/2011/11/Hillier-Sahbaz_An-evidence-based-approach_010408.pdf (accessed 20 August 2021).

Hillier, B., and Shu, S. 2000. Crime and Urban layout: The need for evidence, in Ballintyne, S., Pease, K., and Mclaren, V. (eds.), *Secure Foundations: Key Issues in Crime Prevention, Crime Reduction and Community Safety*. IPPR, London.

HM Treasury. 1984. *Green Booklet*. London.

HM Treasury. 1988. *Policy Evaluation: A Guide for Managers*. London.

HM Treasury. 2003. *The Green Book: Appraisal and Evaluation in Central Government*. London.

Hodkinson, S., and Robbins, G. 2013. The return of class war conservatism? Housing under the UK coalition government. *Critical Social Policy* 33, 1, 57–77.

Hogwood, B., and Gunn, L. 1984. *Policy Analysis for the Real World*. Oxford University Press, Oxford.

Home Office Crime Prevention Centre. 1994. *Police Architectural Liaison Guidance Manual*. Stafford.

Home Office. 2014. *Protecting Crowded Places: Design and Technical Issues*. London. https://www.gov.uk/government/publications/protecting-crowded-places-design-and-technical-issues (accessed 20 August 2021).

Home Office. 2017. *Crowded Places*. London. https://www.gov.uk/government/publications/crowded-places-guidance (accessed 20 August 2021).

Hope, T., and Dowds, L. 1987. The use of local surveys in evaluation research: Examples from community crime prevention *British Criminology Conference* July 1987.

Hope, T., and Foster, J. 1992. Conflicting forces: Changing dynamics of crime and community on a 'problem estate'. *The British Journal of Criminology* 32,4, 488–504.

Horizon: The Writing on the Wall. 1974. Directed by Mansfield, J. M., 21:25 London: BBC. http://genome.ch.bbc.co.uk/schedules/bbctwo/england/1974-02-11#at-21.25 (accessed 20 August 2021).

Housing Development Directorate. 1980. *Priority Estates Project: Upgrading Problem Council Estates*. Department of the Environment, London.

Hoyt, L. 2006. Imported ideas: The transnational transfer of urban revitalization policy. *International Journal of Public Administration* 29, 221–243.

Hubbard, P., and Lees, L. 2018. The right to community: Legal geographies of resistance on London's final gentrification frontier. *CITY* 22, 1, 8–25.

Hunter-Tilney, L. 2012. Architecture: Paradise lost. *New Statesman* 12 March, n.p.

Hunter-Tilney, L. 2019. London rapper Fredo on the neighbourhood that shaped his music, *Financial Times* February, https://www.ft.com/content/5355f564-288a-11e9-88a4-c32129756dd8 (accessed 20 August 2021).

Hyra, D. 2008. *The New Urban Renewal: The Economic Transformation of Harlem and Bronzeville.* University of Chicago Press, Chicago.

Imrie, R., and Raco, M. (eds.). 2003. *Urban Renaissance? New Labour, Community and Urban Policy.* Policy Press, Bristol.

Jacobs, J. 1961. *The Death and Life of Great American Cities.* Random House, New York.

Jacobs, J.M. 2012. Urban geographies 1: Still thinking cities relationally. *Progress in Human Geography* 36, 3, 412–422.

Jacobs, J.M., and Lees, L. 2013. Defensible space on the move: Revisiting the urban geography of Alice Coleman. *International Journal of Urban and Regional Research* 37, 5, 1559–1583.

Jacobs, K., and Manzi, T. 1996. Discourse and policy change: The significance of language for housing research. *Housing Studies* 11, 4, 543–560.

Jacobs, K., and Manzi, T. 2013. Modernisation, marketisation and housing reform: The use of evidence-based policy as a rationality discourse. *People, Place & Policy Online* 7, 1, 1–13.

Jeffrey, C. 1971. *Crime Prevention through Environmental Design.* Sage, Beverly Hills, C.A.

Jeffrey, C. 1999. CPTED: Past, present and future. *4th International CPTED Association Conference.* Canada 20–22 September.

Jenkins, S. 2010. London planners must embrace civilised living. *London Evening Standard* 28 September, n.p.

Jenks, M. 1988. Housing problems and the dangers of certainty, in Teymur, N., Markus, T.A., and Woolley, T. (eds.), *Rehumanizing Housing.* Butterworths, Oxford, pp. 53–60.

Jones, P., and Evans, J. 2008. *Urban Regeneration in the UK: Theory and Practice.* Sage, London.

Kanivk, R. 1991. 'Guinea Pigs' in revolt. *Daily Express* 4 January, n.p.

Kennedy, S. 2016. Urban policy mobilities, argumentation and the case of the model city. *Urban Geography* 37, 96–116.

Khirfan, L., Momani, B., and Jaffer, Z. 2013. Whose authority? Exporting Canadian urban planning expertise to Jordan and Abu Dhabi. *Geoforum* 50, December, 1–9.

Kinder, K. 2016. *DIY Detroit: Making Do in a City without Services.* University of Minnesota Press, Minneapolis.

Kingfisher, C. 2013. *A Policy Travelogue: Tracing Welfare Reform in Aotearoa/New Zealand and Canada.* Berghahn Books, Oxford.

Kitchen, T., and Schneider, R. 2005. Crime and the design of the built environment: Anglo-American comparisons of policy and practice, in Hillier, J. and Rooksby, E. (eds.), *Habitus: A Sense of Place.* Ashgate, Hampshire, pp. 258–282.

Knevitt, C. 1986. Estate crime cut by removing walkways. *The Times* 17 January, n.p.

Knoblauch, J. 2014. The economy of fear: Oscar Newman launches crime prevention through urban design (1969-197x). *Architectural Theory Review* 19, 3, 336–354.

Kong, L., Gibson, C., Khoo, L.-M., and Semple, A.-L. 2006. Knowledges of the creative economy: Towards a relational geography of diffusion and adaptation in Asia. *Asia Pacific Viewpoint* 4, 2, 173–194.

Koolhass, R. 1978. *Delirious New York: A Retroactive Manifesto for Manhattan*. The Monacelli Press, New York.

Kuhn, T. 1970. *The Structure of Scientific Revolutions*. University of Chicago Press, Chicago.

Lancione, M., Stefanizzi, A., and Gaboardi, M. 2017. Passive adaptation or active engagement? The challenges of housing first internationally and in the Italian case. *Housing Studies* 33, 40–57.

Larner, W., and Laurie, N. 2010. Travelling technocrats, embodied knowledges: Globalising privatisation in telecoms and water. *Geoforum* 41, 2, 218–226.

Larner, W., and Le Heron, R. 2002. The spaces and subjects of a globalising economy: A situated exploration of method. *Environment and Planning D* 20, 6, 753–774.

Leary-Owhin, M. 2018. Public Spaces of Hope: Reflections on the gentrification glass ceiling, unpublished paper presented to AESOP 2018 Annual Congress, http://www.academia.edu/36723125/Public_Spaces_of_Hope_Reflections_on_the_Gentrification_Glass_Ceiling (accessed 20 August 2021).

Leather, P., and Nevin, B. 2013. The housing market renewal programme: Origins, outcomes, and the effectiveness of public policy interventions in a volatile market. *Urban Studies* 50, 5, 856–875.

Lees, L. 2003. Visions of 'Urban Renaissance': The urban task force report and the urban white paper, in Imrie, R. and Raco, M. (eds.), *Urban Renaissance? New Labour, Community and Urban Policy*. Policy Press, Bristol, pp. 61–82.

Lees, L. (ed.). 2004. *The Emancipatory City: Paradoxes and Possibilities?* Sage, London.

Lees, L. 2014. The urban injustices of New Labour's 'new urban renewal': The case of the Aylesbury Estate in London. *Antipode* 46, 4, 921–947.

Lees, L., and Ferreri, M. 2016. Resisting gentrification on its final frontiers: Lessons from the Heygate Estate in London (1974–2013). *Cities* 57, 14–24.

Lees, L., and Hubbard, P. 2020. Legal geographies of resistance to gentrification and displacement: Lessons from the Aylesbury Estate in London, in Adey, P., Bowstead, J.C., Brickell, K., Desai, V., Dolton, M., Pinkerton, A., and Siddiqi, A. (eds.), *Handbook of Displacement*. Palgrave, Cham, Switzerland, pp. 753–770.

Lees, L., and Hubbard, P. 2021. 'So, don't you want us here no more?' Slow violence, frustrated hope, and racialised struggle on London's council estates. *Housing, Theory and Society*. (https://doi.org/10.1080/14036096.2021.1959392).

Lees, L., Shin, H., and Lopez-Morales, E. 2016. *Planetary Gentrification*. Polity Press, Cambridge.

Lees, L., and White, H. 2020. The social cleansing of London council estates: Contemporary experiences of 'accumulative dispossession'. *Housing Studies* 35, 10, 1701–1722.

Levitt, D., and McCafferty, J. 2018. *The Housing Design Handbook*. Routledge, London and New York.

Ley, D. 1974a. Review: 'The social construction of communities' – Suttles, Gerald; 'Defensible space' – Newman, Oscar. *Annals of the Association of American Geographers* 64, 1, 156–158.

Ley, D. 1974b. *Black Inner City as Frontier Outpost: Images and Behaviour of a Philadelphia Neighborhood*. Association of American Geographers, Washington DC.

Ley, D. 1977. Social geography and the taken for granted world. *Transactions of the Institute for British Geographers* 2, 478–512.

Lipman, A., and Harris, H. 1988. Dystopian aesthetics – A refusal from 'nowhere', in Teymur, N., Markus, T.A., and Woolley, T. (eds.), *Rehumanizing Housing*. Butterworths, Oxford, pp. 182–194.

London Borough of Tower Hamlets. 2009. *Creating a Safe Environment: Supplementary Planning Guidance 'Design Out' Crime*. London Borough of Tower Hamlets, London.

Lorimer, H. 2003. The geographical field course as active archive. *Cultural Geographies* 10, 3, 278–308.

Lovell, H. 2017. Are policy failures mobile? An investigation of the advanced metering infrastructure program in the State of Victoria, Australia. *Environment and Planning A* 49, 314–331.

Lovell, H. 2019. Policy failure mobilities. *Progress in Human Geography* 43, 1, 46–63.

Lowenfeld, J. 2008. Estate regeneration in practice: The Mozart Estate 1985–2004, in Harwood, E. and Power, A. (eds.), *Housing the Twentieth Century Nation*. The Twentieth Century Society, London, pp. 164–174.

Lowry, S. 1990. Families and flats. *British Medical Journal* 300, 245–247.

Luckman, S., Gibson, C., and Le, T. 2009. Mosquitoes in the mix: How transferable is creative city thinking? *Singapore Journal of Tropical Geography* 30, 1, 70–85.

Lynch, K. 1960. *The Image of the City*. MIT Press, Cambridge, MA.

Lynch, K. 1981. *A Theory of Good City Form*. MIT Press, Cambridge, MA.

Macaskill, H. 1985. Best Seller! The difficulties of publishing a book from scratch. *The Guardian* 28 October, n.p.

Macaskill, H. 1986. *List of Press Articles on Utopia on Trial*. (Archive catalogue). Hilary Shipman Ltd, London.

Madden, D., and Marcuse, P. 2016. *In Defense of Housing: The Politics of Crisis*. Verso, London.

Maddrell, A. 2009. *Complex Locations: Women's Geographical Work in the UK 1850–1970*. Wiley-Blackwell, Oxford.

Malpass, P. 1988. Utopia in context: State, class and the restructuring if the housing market in the twentieth century, in Teymur, N., Markus, T.A., and Woolley, T. (eds.), *Rehumanizing Housing*. Butterworths, Oxford, pp. 142–153.

Malpass, P. 2005. *Housing and the Welfare State: The Development of Housing Policy in England*. Palgrave Macmillan, Basingstoke.

Mansfield, J., Dir. 1974. *Horizon: The Writing on the Wall*. BBC documentary.

Markus, T. 1988. Rehumanizing the dehumanized, in Teymur, N., Markus, T.A., and Woolley, T. (eds.), *Rehumanizing Housing*. Butterworths, Oxford, pp. 1–16.

Marsh, D., and McConnell, A. 2010. Towards a framework for establishing policy success. *Public Administration* 88, 2, 564–583.

Marshall, S. 2012. Science, pseudo-science and urban design. *Urban Design International* 17, 4, 257–271.

Mawby, R. 1977. Defensible Space: A theoretical and empirical appraisal. *Urban Studies* 14, 169–179.

Mawson, J., Beazley, M., Burfitt, A. 1995. *The Single Regeneration Budget – The Stocktake*. Centre for Urban and Regional Studies, University of Birmingham, Birmingham.

Mayhew, P., Clarke, R.V., Hough, M., and Sturman, A. 1976. *Crime as Opportunity*. *Home Office Research Study No. 34*. Home Office, London.

McCann, E. 2008. Expertise, truth, and urban policy mobilities: Global circuits of knowledge in the development of Vancouver. Canada's 'four pillar' drug strategy. *Environment and Planning A* 40, 4, 885–904.

McCann, E. 2011a. Urban policy mobilities and global circuits of knowledge: Toward a research agenda. *Annals of the Association of American Geographers* 101, 1, 107–130.

McCann, E. 2011b. Veritable inventions: Cities, policies, and assemblage. *Area* 43, 143–147.

McCann, E. 2013. Policy boosterism, policy mobilities, and the extrospective city. *Urban Geography* 34, 5–29.

McCann, E., and Temenos, C. 2015. Mobilizing drug consumption rooms: Inter-place networks and harm reduction drug policy. *Health & Place* 31, 216–223.

McCann, E.J., and Ward, K. (eds.). 2011. *Mobile Urbanism: Cities and Policy Making in the Global Age*. Minneapolis and London, University of Minesota Press.

McCann, E., and Ward, K. 2013. A multi-disciplinary approach to policy transfer research: Geographies, assemblages, mobilities and mutations. *Policy Studies* 34, 1, 2–18.

McCormack, D. 2003. An event of geographical ethics in spaces of affect. *Transactions of the Institute of British Geographers* 28, 4, 488–507.

McDowell, L. 1997. *Capital Culture: Gender at Work in the City*. John Wiley and Sons, Chichester.

McFarlane, C. 2011. *Learning the City: Knowledge and Translocal Assemblage*. Wiley-Blackwell, Chichester.

McGuirk, P. 2016. Practicing fast policy at and beyond the edges of neoliberalism. *Political Geography* 53, 93–95.

McKean, J. 1973. Defend your space from vandals. *Architects' Journal* 158, 47, 1239–1243.

McLaughlin, E., and Muncie, J. (eds.). 1996. *Controlling Crime*. Sage, London.

McMenzie, L., Cook, I., and Laing, M. 2019. Criminological policy mobilities and sex work: Understanding the movement of the 'Swedish model' to Northern Ireland, *The British Journal of Criminology*, https://doi.org/10.1093/bjc/azy058 (accessed 20 August 2021).

Meek, J. 2014. Where will we live? *London Review of Books* 31, 7–16. http://www.lrb. co.uk/v36/n01/james-meek/where-will-we-live (accessed 20 August 2021).

Merry, S. 1981. Defensible space undefended: Social factors in crime control through environmental design. *Urban Affairs Quarterly* 16, 4, 397–422.

Merton, R. 1967. On sociological theories of the middle-range, in Merton, R. (ed.), *On Theoretical Sociology: Five Essays Old and New*. Free Press, New York, pp. 39–72.

Meyer, W., and Guss, D. 2017. *Neo-environmental Determinism: Geographical Critiques*. Palgrave, Basingstoke.

MHCLG. 2018. *The Hackitt Review: Independent Review of Building Regulations and Fire Safety: Final Report*. Ministry of Housing, Communities and Local Government, London. https://www.gov.uk/government/publications/independent-review-of-building-regulations-and-fire-safety-final-report (accessed 20 August 2021).

MHCLG. 2019a. *National Design Guide – Planning Practice Guidance for Beautiful, Enduring and Successful Places*. Ministry of Housing, Communities and Local Government, London. https://www.gov.uk/government/publications/national-design-guide (accessed 20 August 2021).

MHCLG. 2019b. *National Design Guide – Planning Practice Guidance for Beautiful, Enduring and Successful Places*. Ministry of Housing, Communities and Local Government, London. https://www.gov.uk/government/publications/national-design-guide (accessed 20 August 2021).

MHCLG. 2021. National Planning Policy Framework: Draft text for consultation. https://assets.publishing.service.gov.uk/government/uploads/system/uploads/attachment_data/file/961769/Draft_NPPF_for_consultation.pdf (accessed 20 August 2021).

Minkman, E., Van Buuren, M., and Bekkers, V. 2018. Policy transfer routes: An evidence-based conceptual model to explain policy adoption. *Policy Studies* 39, 2, 222–250.

Minton, A. 2009. *Ground Control: Fear and Happiness in the Twenty-first-century City*. Penguin, London.

Minton, A. 2013. The reconfiguration of London is akin to social cleansing. *The Guardian* 27 March.

Mitchell, D. 2000. *Cultural Geography. A Critical Introduction*. Blackwell, Oxford.

Montero, S. 2017. Study tours and inter-city policy learning: Mobilizing Bogota's transportation policies in Guadalajara. *Environment and Planning A* 49, 2, 332–350.

Moran, R., and Dolphin, C. 1986. The defensible space concept: Theoretical and operational exploration. *Environment and Behavior* 18, 3, 396.

Morphet, J. 2014. *How Europe Shapes British Public Policy*. Policy Press, Bristol.

Müller, M. 2015. (Im-)mobile policies: Why sustainability went wrong in the 2014 Olympics in Sochi. *European Urban and Regional Studies* 22, 2, 191–209.

Mumford, L. 1938. *The Culture of Cities*. Harcourt Brace and Company, Florida.

Municipal Journal 1990. Cash boost. 9 March, n.p.

Murie, A. 1997. Linking housing changes to crime. *Social Policy and Administration* 31, 5, 22–36.

Newburn, T., Jones, T., and Blaustein, J. 2018. Policy mobilities and comparative penalty. *Theoretical Criminology* 22, 4, 563–581.

New Society. 1986. Crime reduction possible. 12 March, n.p.

New Statesman. 2012. Architecture: Paradise lost. https://www.newstatesman.com/art/2012/03/coleman-thatcher-interview, (accessed 20 August 2021).

Newman, J., and Clarke, J. 2009. *Publics, Politics and Power.* Sage, London.

Newman, O. 1972. *Defensible Space. Crime Prevention Through Urban Design.* Macmillan, New York.

Newman, O. 1973. *Defensible Space. People and Design in the Violent City,* 2nd edn. Architectural Press, London.

Newman, O. 1976. *Design Guidelines for Creating Defensible Space.* National Institute of Law Enforcement and Criminal Justice Law Enforcement Assistance Administration, U.S. Department of Justice, Washington, DC.

Newman, O. 1980. *Community of Interest.* Anchor Press/Doubleday.

Newman, O. 1995. Defensible space: A new planning tool for urban revitalization. *Journal of the American Planning Association* 61, 2, 14.

NHBC Foundation and Homes and Communities Agency (NHBC & HCA). 2015. *Tenure Integration in Housing Developments: A Literature Review (NF66).* NHBC Foundation, Milton Keynes. https://www.nhbcfoundation.org/publication/tenure-integration-in-housing-developments (accessed 20 August 2021).

NHTPC (National Housing and Town Planning Council). n.d. *High Rise Housing,* 1–45, No longer available.

Nuttall, S. 1988. Review: *Rehumanizing housing. High Rise Quarterly,* 21–23.

Observer. 1994. Yard to probe council payouts. 19 June, 6.

ODPM. 2005. *Planning Policy Statement 1: Delivering Sustainable Development.* Office of the Deputy Prime Minister, London.

ODPM and Home Office. 2004. *Safer Places: The Planning System and Crime Prevention.* Office of the Deputy Prime Minister, London.

Ortolano, G. 2019. *Thatcher's Progress: From Social Democracy to Market Liberalism through an English New Town.* Cambridge University Press, Cambridge.

Paddington Times. 1977. Tenants press to move from prize estate. 28 January, n.p.

Palfrey, C., Thomas, P., and Phillips, C. 2012. *Evaluation for the Real World: The Impact of Evidence in Policy Making.* Bristol University Press, Bristol.

Parsons, W. 2002. From muddling through to muddling up. Evidence based policymaking and the modernisation of British government. *Public Policy and Administration* 17, 3, 43–60.

Pasco, T. 1992. *Secured By Design – A crime prevention philosophy.* MSc dissertation, Cranfield Institute of Technology, UK.

Pasco, T. 1999. *Evaluation of Secured by Design in Public Sector Housing – Final Report.* BRE, Watford.

Pawley, M. 1985. Why the market likes frills. *The Guardian,* 1 July, 8.

Pawson, R. 2013. *The Science of Evaluation: A Realist Manifesto.* Sage, London.

Pawson, R., and Tilly, N. 1997. *Realistic Evaluation*. Sage, London.

Peabody Housing Trust. 2018. *Peabody Design Guide*. Peabody Housing Trust, London. https://www.peabody.org.uk/media/13910/peabody-design-guide.pdf (accessed 20 August 2021).

Peach, C., Robinson, V., and Smith, S. (eds). 1981. *Ethnic Segregation in Cities*. Croom Helm, London.

Pearman, H. 2016. Fault lines. *RIBA Journal*, February, 67. https://www.ribaj.com/culture/fault-lines (accessed 21 August 2021).

Pease, K., and Gill, M. 2011. *Home Security and Place Design: Some Evidence and Its Policy Implications*. Perpetuity Research & Consultancy International, Tunbridge Wells.

Peck, J. 2002. Political economies of scale: Fast policy, interscalar relations, and neoliberal workfare. *Economic Geography* 78, 3, 331–360.

Peck, J. 2005. Struggling with the creative class. *International Journal of Urban and Regional Studies* 29, 4, 740–770.

Peck, J. 2011a. Geographies of policy: From transfer-diffusion to mobility-mutation. *Progress in Human Geography* 35, 773–797.

Peck, J. 2011b. Global policy models, globalizing poverty management: International convergence or fast-policy integration? *Geography Compass* 5, 4, 165–181.

Peck, J. 2012. Austerity urbanism. *City* 16, 6, 626–655.

Peck, J., and Theodore, N. 2010a. Mobilizing policy: Models, methods, and mutations. Themed issue: Mobilizing policy. *Geoforum* 41, 2, 169–174.

Peck, J., and Theodore, N. 2010b. Recombinant workfare, across the Americas: Transnationalizing fast welfare policy. *Geoforum* 41, 2, 195–208.

Peck, J., and Theodore, N. 2012. Follow the policy: A distended case approach. *Environment and Planning A* 44, 1, 21–30.

Phillips, J. 2006. Agencement/assemblage. *Theory, Culture and Society* 23, 108–109.

Pinch, S. 1998. Knowledge communities, spatial theory and social policy. *Social Policy and Administration* 32, 556–571.

Pinto, R. 1991. The impact of estate action on developments in council housing, management and effectiveness. Unpublished PhD thesis, London School of Economics.

Pinto, R. 1993. An analysis of the impact of estate action schemes. *Local Government Studies* 19, 1, 37–55.

Platt, S. 1989. Power to the people. *Roof*, September-October, 24–25.

Pollitt, C. 1993. Occasional excursions. A brief history of policy evaluation in the UK. *Parliamentary Affairs* 46, 3, 353–363.

Popkin, S., Katz, B., Cunningham, M., Brown, K., Gustafson, J., and Turner, M. 2004. *A Decade of HOPE VI: Research Findings and Policy Challenges*. The Urban Institute/The Brookings Institute, Washington, DC.

Popper, K. 1963. *Conjectures and Refutations: The Growth of Scientific Knowledge in the Sciences of Wealth and Society*. Routledge and Kegan Paul, London.

Power, A. 1982. *Priority Estates Project 1982: Improving Problem Council Housing Estates*. DoE, London.

Power, A. 1984a. *Local Housing Management – A Priority Estates Project Survey.* DoE, London.

Power, A. 1984b. Rescuing unpopular council estates through local management. *The Geographical Journal* 150, 3, 359–362.

Power, A. 1985. The development of unpopular council housing estates and attempted remedies 1895–1984. PhD thesis, London School of Economics.

Power, A. 1986. Home is home. *Architects' Journal* 187, 43, 30–31.

Power, A. 1987. *PEP Guide to Local Housing Management.* DoE, London.

Power, A. 1998. *Estates on the Edge: The Social Consequences of Mass Housing in Northern Europe.* Palgrave Macmillan, Basingstoke.

Power, A., and Tunstall, R. 1995. *Swimming against the Tide: Polarisation or Progress on 20 Unpopular Council Estates, 1980–1995.* Joseph Rowntree Foundation, York.

Poyner, B. 1983. *Design against Crime – Beyond Defensible Space.* Butterworths, Oxford.

Poyner, B. 1986. Lessons from Lisson Green: An evaluation of walkway demolition on a British Housing Estate, in Clarke, R.V. (ed.), *Crime Prevention Studies*, Vol. 3. Lynne Reinner, London., https://popcenter.asu.edu/sites/default/files/05_poyner.pdf (accessed 20 August 2021).

Price Waterhouse. 1995. *Evaluation of Design Improvement Controlled Experiment (DICE); Report on Rogers Estate, Globe Town.* London.

Price Waterhouse. 1996. *Evaluation of Design Improvement Controlled Experiment (DICE); Report on Ranwell East Estate, Bow, Final Report.* London.

Price Waterhouse. 1997a. *The Design Improvement Controlled Experiment (DICE): An Evaluation of the Impact, Costs, and Benefits of Estate Re-modelling.* DoE, London.

Price Waterhouse. 1997b. *An Evaluation of DICE (Design Improvement Controlled Experiment) Schemes: Regeneration Research Summary 11.* DoE, London.

Queen's Park Forum. 2008. *Queen's Park Neighbourhood Plan 2008–12.* Queen's Park Forum, London.

Raines, H. 1988. Defying Tradition: Prince Charles recasts his role. *The New York Times* 21 February, n.p.

Randall, R. 1998. Letter: Criminal Analysis. *Building Design* 1423, 20 November, 10.

Ravetz, A. 1986. Review 'Utopia on Trial'. *Planning Perspectives* 1, 279–296.

Ravetz, A. 1988. Malaise, design and history: Scholarship and experience on trial, in Teymur, N., Markus, T.A., and Woolley, T. (eds.), *Rehumanizing Housing.* Butterworths, Oxford, 154–165.

Raymen, T. 2015. Designing-in crime by designing out the social? Situational crime prevention and the intensification of harmful subjectivities. *The British Journal of Criminology* 56, 497–514.

Reynald, D., and Elffers, H. 2009. The future of Newman's Defensible Space Theory: Linking defensible space and the routine activities of place. *European Journal of Criminology* 6, 25, 25–46.

Reynolds, I., Nicholson, C., Crowther, S., Birley, R., and Bell, A. 1972. *The Estate outside the Dwelling. Reactions of Residents to Aspects of Housing Layout.* DoE, London.

Roberts, M. 1988. Caretaking – who cares?, in Teymur, N., Markus, T.A., and Woolley, T. (eds.), *Rehumanizing Housing*. Butterworths, Oxford, 123–132.

Robinson, J. 2015. 'Arriving at' urban policies: The topological spaces of urban policy mobility. *International Journal of Urban and Regional Research* 39, 4, 831–834.

Robinson, J. 2018. 'Arriving at' urban policies/the urban: Traces of elsewhere in making city futures, in Soderstrom, O., Randeria, S., Ruedin, D., D'Amato, G., and Panese, F. (eds.), *Critical Mobilities*. EPFL Press, Routledge, pp. 1–28.

Rogers, R., and Power, A. 2000. *Cities for a Small Planet*. London, Faber and Faber.

ROOF. 1989. September/October, p. 24. No longer available.

Rose, G. 1993. *Feminism and Geography: The Limits of Geographical Knowledge*. Polity Press, Cambridge.

Rose, R. 1991. What is lesson-drawing? *Journal of Public Policy* 11, 3–30.

Rose, R. 1993. *Lesson-Drawing in Public Policy*. Chatham House Publishers, Chatham, NJ.

Rowlands, R., Murie, A., and Tice, A. 2006. *More Than Tenure Mix: Developer and Purchaser Attitudes to New Housing Estates*. JRF/CIH, York.

Rubenstein, H. 1980. *The Link between Crime and the Built Environment. The Current State of Knowledge*. National Institute of Justice, Washington DC.

Rudlin, D. 2015. Feeling safe. *Here and Now: Academy of Urbanism Journal* 5, Spring, 37–39.

Rusu, A., and Löblová, O. 2019. Failure is an option: Epistemic communities and the circulation of Health Technology Assessment, in Baker, T. and Walker, C. (eds.), *Public Policy Circulation: Arenas, Agents and Actions*. Edward Elgar, London, pp. 103–120.

Rycroft, S., and Cosgrove, D. 1995. Mapping the modern nation: Dudley stamp and the land utilisation survey of Britain. *History Workshop Journal* 40, 1, 91–105.

Rydin, Y. 2003. *Conflict, Consensus and Rationality in Environmental Planning: An Institutional Discourse Approach*. Oxford University Press, Oxford.

Sampson, R. 2010. Eliding the theory/research and basic/applied divides: Implications of Merton's 'middle range', in Calhoun, C. (ed.), *Robert K. Merton: Sociology of Science and Sociology as Science*. Columbia University Press, New York, pp. 63–78.

Sampson, R., Raudenbush, S., and Earls, F. 1997. Neighborhoods and violent crime: A multilevel study of collective efficacy. *Science and Public Affairs* 277, 918–924.

Savills. 2016. Completing London's Streets. How the regeneration and intensification of housing estates could increase London's supply of homes and benefit residents, Report to the Cabinet Office. https://pdf.euro.savills.co.uk/uk/residential—other/completing-london-s-streets-080116.pdf (accessed 20 August 2021).

Savills. 2018. This is Manchester … as you've never seen it before. https://www.placenorthwest.co.uk/insight/this-is-manchester-as-youve-never-seen-it-before (accessed 20 August 2021).

Sayer, A. 1991. Behind the locality debate: Deconstructing geography's dualisms. *Environment and Planning A* 32, 2, 283–308.

Scanlon, K., Whitehead, C., Sagor, E., and Mossa, A. 2016. *New London Villages: Creating Community.* LSE Consulting, Berkeley Group, London, UK.

Schmidt, V., and Thatcher, M. 2013. Theorizing ideational continuity: The resilience of neo-liberal ideas in Europe, in Schmidt, V. and Thatcher, M. (eds.), *Resilient Liberalism in Europe's Political Economy.* Cambridge University Press, Cambridge, UK, pp. 1–50.

Schneider, R., and Kitchen, T. 2002. *Planning for Crime Prevention: A Transatlantic Perspective.* Routledge, London and New York.

Schneider, R., and Kitchen, T. 2007. *Crime Prevention and the Built Environment.* Routledge, London and New York.

Schön, D. 1983. *The Reflective Practitioner: How Professionals Think in Action.* Basic Books, New York.

Scott, J. 1998. Letter: Down the line. *Building Design* 1423, 20 November, 10.

Secured by Design Principles (2004) http://archive.wyreforestdc.gov.uk/media/107729/EB077-SBD-principles.pdf (accessed 21 August 2021).

Sennett, R. 1986. *The Fall of Public Man.* Faber and Faber, London.

SEU. 1998. *Bringing Britain Together: A National Strategy for Neighbourhood Renewal.* Social Exclusion Unit, HMSO, London.

Severs, D. 2010. Rookeries and no-go estates: St. Giles and Broadwater Farm, or middle-class fear of 'non-street' housing. *The Journal of Architecture* 15, 4, 449–497.

Shaftoe, H. 2004. *Crime Prevention: Facts, Fallacies and the Future.* Palgrave, Basingstoke.

Sherratt, N., Goldblatt, D., Mackintosh, M., and Woodward, K. 2000. *An Introduction to the Social Sciences: Understanding Social Change, Workbook 1.* The Open University, Milton Keynes.

Silver, P. 1995. *Case Study: RIBA Part III Examination in Professional Practice.* University College, London.

Simmons, B., Dobbin, F., and Garrett, G. 2006. Introduction: The international diffusion of liberalism. *International Organization* 60, 4, 781–810.

Simmons, B., and Elkins, Z. 2004. The globalization of liberalization: Policy diffusion in the international political economy. *American Political Science Review* 98, 1, 171–189.

Slater, T. 2018. The invention of the 'sink estate': Consequential categorization and the UK housing crisis, in Tyler, I. and Slater, T. (eds.), *The Sociology of Stigma.* Sage: Sociological Review Monograph Series, London.

Smith, N. 1996. *The New Urban Frontier: Gentrification and the Revanchist City.* Routledge, London and New York.

Smith, S. 1986a. Review 'Utopia on Trial. Vision and Reality in Planned Housing'. *Urban Studies* 23, 3, 224–246.

Smith, S. 1986b. *Crime, Space and Society.* Cambridge University Press, Cambridge.

Smith, S. 1987. Design against crime? Beyond the rhetoric of residential crime prevention. *Journal of Property Management* 5, 146–150.

Smith, S. 2003. Reply to comments on 'Crime, space and society' in 'Classics in Human Geography Revisited'. *Progress in Human Geography* 27, 337–339.

SNU. 1988a. *Report of the Safe Neighbourhoods Unit's Work 1981–86*. Safe Neighbourhoods Unit, London.

SNU. 1988b. *The Mozart Survey Part 1 – A Study of Design Modification and Housing Management Innovation*. Safe Neighbourhoods Unit, London.

SNU. 1993. *Crime Prevention on Council Estates*. Department of the Environment, London.

SNU. 2009. *Review of Safe Neighbourhoods Unit's Work on Housing Estates 1981–2009*. Safe Neighbourhoods Unit, London.

Soaita, A. 2018. *Mapping the literature of 'policy transfer' and housing*, Working Paper, UK Collaborative Centre for Housing Evidence.

Soaita, A., Marsh, A., and Gibb, K. 2021. Policy movement in housing research: A critical interpretative synthesis. *Housing Studies*. https://www.tandfonline.com/doi/full/10.1080/02673037.2021.1879999 (accessed 20 August 2021).

Solomos, J. 1988. *Race and Racism in Britain*. Macmillan, Basingstoke.

Spicker, P. 1987. Poverty and Depressed Estates: A critique of 'Utopia on Trial'. *Housing Studies* 2, 4, 283–292.

Spring, M. 1994. Crime guru in Crack City deal. *Building* 259, 11, 10.

Spring, M. 1997. Alice on trial. *Building* 262, 44, 46–50.

Stein, C., Boris, M., Glaze, G., and Putz, R. 2015. Learning from failed policy mobilities: Contradictions, resistances and unintended outcomes in the transfer of 'Business Improvement Districts' to Germany. *European Urban and Regional Studies* 24, 1, 35–49.

Stone, D. 1999. Learning lessons and transferring policy across time, space and disciplines. *Politics* 19, 51–59.

Stone, D. 2004. Transfer agents and global networks in the 'transnationalization' of policy. *Journal of European Public Policy* 11, 3, 545–566.

Stone, D. 2012. Transfer and translation of policy. *Policy Studies* 33, 6, 483–499.

Stone, D. 2016. Understanding the transfer of policy failure: Bricolage, experimentalism and translation. *Policy & Politics* 45, 1, 55–70.

Strategic Policy Making Team. 1999. *Professional Policy Making for the Twenty First Century*. Cabinet Office, London.

Sturman, A., and Wilson, S. 1976. Vandalism research aimed at specific remedies. *Municipal Engineering* 7 May, 703–713.

Suchman, E. 1967. *Evaluative Research: Principles in Public Service and Action Programs*. Russell Sage, New York.

Swenarton, M. 2017. *Cook's Camden: The Making of Modern Housing*. Lund Humphries Publishers Ltd., London

Sykes, J. (ed.). 1979. *Designing against Vandalism*. The Design Council, London.

Tait, M., and Jensen, O. 2007. Travelling ideas, power and place: The cases of urban villages and business improvement districts. *International Planning Studies* 12, 2, 107–128.

Taylor (Lord). 2012. *External Review of Government Planning Practice Guidance*. Department for Communities and Local Government, London. https://www.gov.uk/

government/uploads/system/uploads/attachment_data/file/39821/taylor_review.pdf (accessed 20 August 2021).

Taylor, L. 1973. The meaning of the environment, in Ward, C. (ed.), *Vandalism. The Architecture Press*, London, pp. 54–63.

Temenos, C., and Baker, T. 2015. Enriching urban policy mobilities research. *International Journal of Urban and Regional Research* 39, 4, 841–843.

Teymur, N., Markus, T., and Woolley, T. (eds.). 1988. *Rehumanizing Housing*. Butterworths, Oxford.

Thatcher, M. 1993. *The Downing Street Years*. Harper Collins, London.

Theodore, N. 2019. Policy mobilities, in *Oxford Bibliographies*. https://www.oxford bibliogrphies.com/view/document/obo-9780199874002/obo-9780199874002-0205.xml (accessed 20 August 2021).

Thompson, M. 2020. *Reconstructing Public Housing: Liverpool's Hidden History of Collective Alternatives*. Liverpool University Press, Liverpool.

Tiesdell, S. 2001. A forgotten policy, a perspective on the evolution and transformation of housing action trust policy. *European Journal of Housing Policy* 1, 3.

Till, J. 1998. Architecture of the impure community, in Hill, J. (ed.), *Occupations of Architecture*. Routledge, London, pp. 34–42.

Towers, G. 2000. *Shelter Is Not Enough: Transforming Multi-storey Housing*. The Policy Press, Bristol.

Tunstall, R. 2020. *The Fall and Rise of Social Housing: 100 Years on 20 Estates*. Policy Press, Bristol.

Vale, L. 2013. Public housing in the United States: Neighbourhood renewal and the poor, in Carmon, N. and Fainstein, S. (eds.), *Policy, Planning and People: Promoting Justice in Urban Development*. University of Pennsylvania Press, pp. 285–306.

Vallet, B. 2006. Aux origines de l'espace défendable: Une critique de l'urban renewal. *Les Cahiers De La Sécurité* 59, 4, 235–254.

Voluntary Housing. 1992. *Voluntary Housing*, p. 13. No longer available.

Wajcman, J. 1999. *Managing like a Man: Women and Men in Corporate Management*. Allen and Unwin.

Walker, C. 2018. Policy transfer in a corporatist context: Agents, adjustments and continued innovation. *Public Policy and Administration* 34, 3, 308–328.

Wang, J. 2004. The global reach of a new discourse: How far can 'creative industries' travel? *International Journal of Cultural Studies* 7, 1, 9–19.

Ward, C. 1973a. Review of 'Defensible Space'. *Architects' Journal* 158, 47, 1243.

Ward, C. (ed.). 1973b. *Vandalism*. London: Architectural Press.

Ward, C. 1976. *Housing an Anarchist Approach*. Freedom Press, London.

Ward, K. 2006. 'Policies in motion', urban management and state restructuring: The trans-local expansion of business improvement districts. *International Journal of Urban and Regional Research* 30, 54–75.

Ward, K. 2007. Business improvement districts: Policy origins, mobile policies and urban liveability. *Geography Compass* 1, 3, 657–672.

Ward, K. 2011. Policies in motion and in place: The case of the business improvement districts, in McCann, E. and Ward, K. (eds.), *Mobile Urbanism: Cities and Policy-making in a Global Age*. Minnesota University Press, Minneapolis, pp. 71–96.

Ward, S. 2010. What did the Germans ever do for us? A century of British learning about and imagining modern town planning. *Planning Perspectives* 25, 117–140.

Warman, C. 1991. Facelift for written off estates. *The Times* 16 January n.p.

Warwick, E. 2015. Policy to reality: Evaluating the evidence trajectory for English eco-towns. *Building Research and Information* 43, 486–498.

Warwick, E., and Lees, L. in review. Osmosis across 'Defensible Space': The diffusion and concentration of activities around housing estates. *Urban Geography*.

Wates, N., and Knevitt, C. 1987. *Community Architecture*. Penguin, London.

Watt, P., and Minton, A. 2016. London's housing crisis and its activisms - Introduction. *CITY* 20, 2, 204–221.

Webber, S. 2015. Mobile adaptation and sticky experiments: Circulating best practices and lessons learned in climate change adaptation. *Geographical Research* 53, 1, 26–38.

Wells, K. 2019. Policy failing: A repealed right to shelter. *Urban Geography* 41, 9, 1139–1157.

Wells, P. 2007. New Labour and evidence based policy making: 1997–2007. *People, Place and Policy Online* 1, 22–29. https://extra.shu.ac.uk/ppp-online/new-labour-and-evidence-based-policy-making-1997-2007/ (accessed 20 August 2021).

Wenger, E. 1998. *Communities of Practice: Learning, Meaning and Identity*. Cambridge University Press, Cambridge.

White, T., and Serin, B. 2021. *High-rise Residential Development: An International Evidence Review*. UK Collaborative Centre for Housing Evidence. https://housingevidence. ac.uk/high-rise-residential-development-an-international-evidence-review/ (accessed 20 August 2021).

Whyte, W. 1956. *The Organisation Man*. Doubleday, Garden City.

Williams, P. 1985. Alice in Wonderland. *Housing* 18–19.

Williams, S. 2014. *Stephen Williams Announces Plans to Simplify Housing Standards*. DCLG, London. https://www.gov.uk/government/news/stephen-williams-announces-plans-to-raise-housing-standards (accessed 20 August 2021).

Wilson, J., and Kelling, G. 1982. Broken windows. The police and neighbourhood safety. *The Atlantic Monthly* 211, 3, 29–38.

Wilson, S. 1978a. Updating defensible space. *Architects' Journal* 11 October, 674.

Wilson, S. 1978b. Vandalism and defensible space on London housing estates, in Clarke, R.G.V. (ed.), *Tackling Vandalism Home Office Research Study 47*. Home Office Research Unit, London.

Wilson, S. 1979. Observations on the nature of vandalism, in Sykes, J. (ed.), *Designing against Vandalism*. The Design Council, London.

Wilson, S. 1980. Vandalism and 'defensible space' on London housing estates, in Clarke, R.G.V. and Mayhew, P. (eds.), *Designing Out Crime*. Home Office Research Unit, London.

Wilson, S. 1981a. A new look at Newman. *RIBA Journal* 44, May, 50–51.

Wilson, S. 1981b. Design against crime. *Building Design* 538, 27 March, 38.

Wolman, H. 1992. Understanding cross national policy transfers: The case of Britain and the US. *Governance* 5, 27–45.

Wright, P. 2009. *A Journey through Ruins: The Last Days of London*. Oxford University Press, Oxford.

Yancey, W. 1971. Architecture, interaction, and social control: The case of a large-scale public housing project. *Environment and Behavior* 3, 1, 3–21.

Young, K., Ashby, D., Boaz, A., and Grayson, L. 2002. Social Science and the evidence-based policy movement. *Social Policy and Society* 1, 3, 215–224.

Zipp, S. 2010. *Manhattan Projects: The Rise and Fall of Urban Renewal in Cold War New York*. Oxford University Press, USA.

Index